GCSE
Success

REVISION GUIDE

Chemistry

Emma Poole

Contents

Revised

A copy of the periodic table can be found on the inside back cover.

Atomic Structure

Structure of the Atom

An **atom** has a very small, central **nucleus** that is surrounded by shells of **electrons**. The nucleus is found at the centre of the atom. It contains **protons** and **neutrons**:

- Protons have a mass of 1 **atomic mass unit** (**amu**) and a charge of 1+.
- Neutrons also have a mass of 1 amu, but no charge.
- Electrons have a negligible mass and a charge of 1−.

All atoms are neutral, therefore there is no overall charge so the number of protons must be equal to the number of electrons.

The **mass number** is the number of protons added to the number of neutrons. The **atomic number** is the number of protons (so it is also known as the **proton number**). All the atoms of a particular element have the same number of protons, for example carbon atoms always have six protons. Atoms of different elements have different atomic numbers.

Sodium has an atomic number of 11, so every sodium atom has 11 protons.

A sodium atom has no overall charge, so the number of electrons must be the same as the number of protons. Sodium atoms, therefore, have 11 electrons.

The number of neutrons is given by the mass number minus the atomic number. In sodium that is 23 − 11 = 12 neutrons.

Mass number and atomic number

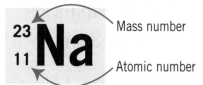

Electrons occupy the lowest available **shell** (or energy level). This is the shell that is closest to the nucleus. When this is full the electrons start to fill the second shell and so on. The first shell may contain up to two electrons, while the second shell may contain up to eight electrons.

The electron structure of an atom is important because it determines how the atom (and, therefore, the element) will react.

Structure of an atom

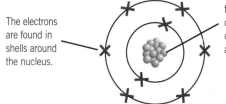

The electrons are found in shells around the nucleus.

The nucleus is found at the centre of the atom and contains neutrons and protons.

✓ Maximise Your Marks

To get an A*, you need to be able to recall the radius and mass of a typical atom.

Atoms are very small. They have a radius of about 10^{-10} m and a mass of about 10^{-23} g.

Elements

A substance that is made of atoms with the same atomic number is called an **element**. Elements cannot be broken down chemically. Atoms of different elements have different properties. About 100 different elements have been discovered. The elements can be represented by **symbols**. Approximately 80 per cent of the elements are **metals**. Metals are found on the left-hand side and in the centre of the periodic table. The **non-metal** elements are found on the right-hand side of the periodic table. Elements with **intermediate properties**, such as germanium, are found in group 4.

Build Your Understanding

Elements in the same group of the periodic table have similar chemical properties because they have the same number of electrons in their outer shells.

Magnesium

- Number of protons = 12
- Number of electrons = 12
- Electronic structure = 2, 8, 2

Magnesium is in group 2 of the periodic table because it has two electrons in its outer shell. It is in period 3 of the periodic table because it has three shells of electrons.

Across a period each element has one extra proton in its nucleus and one extra electron in its outer shell of electrons. This means an electron shell is filled with electrons across a period.

Magnesium is in group 2

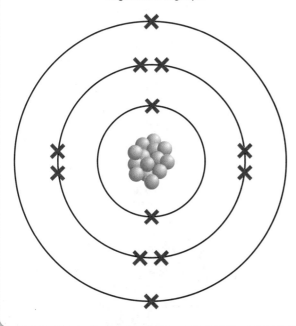

Isotopes

Isotopes of an element have the same number of protons, but a different number of neutrons. So, they have the same atomic number but a different mass number.

The Isotopes of Chlorine

Chlorine has two isotopes:

- 17 protons
- 17 electrons
- 18 neutrons

$^{35}_{17}\text{Cl}$

- 17 protons
- 17 electrons
- 20 neutrons

$^{37}_{17}\text{Cl}$

These isotopes will have slightly different physical properties, but will react identically in chemical reactions because they have identical numbers of electrons.

Learn the definition for relative atomic mass.

The relative atomic mass of an element compares the mass of atoms of the element with the carbon-12 isotope. The existence of isotopes means that some elements have relative atomic masses that are not a whole number, for example chlorine has a relative atomic mass of 35.5:

- 25 per cent of chlorine atoms have an atomic mass of 37.
- 75 per cent of chlorine atoms have an atomic mass of 35.
- This gives an average atomic mass of 35.5.

❓ Test Yourself

1. What does the nucleus of an atom contain?
2. Which particles are found in shells around the nucleus?
3. What is the charge and mass of a proton?
4. What is the charge and mass of an electron?

⭐ Stretch Yourself

1. Calcium and magnesium both belong to group 2 of the periodic table. Why does the element calcium react in a similar way to the element magnesium?

Atoms and the Periodic Table

The History of the Atom

Ideas about atoms have changed over time as more evidence has become available. Scientists look at the evidence that is available and use this to build a model of what they think is happening.

As new evidence emerges they re-evaluate the model. If the model fits with the new evidence they keep it. If the model no longer works they change it.

Build Your Understanding

John Dalton

In the early 1800s, John Dalton developed a theory about atoms, which included these predictions:
- Elements are made up of small particles called atoms.
- Atoms cannot be divided into simpler substances.
- All atoms of the same element are the same.
- Atoms of each element are different from atoms of other elements.

J.J. Thomson

Between 1897 and 1906 Thomson discovered that atoms could be split into smaller particles. He discovered electrons and found that they:
- Have a negative charge.
- Are very small.
- Are deflected by magnetic and electric fields.

He thought that atoms consisted of tiny negative electrons surrounded by a 'sea' of positive charge. Overall, the atom was neutral. This was called the plum-pudding theory of atoms.

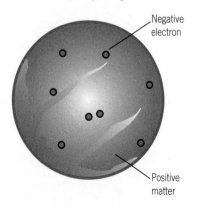

J.J. Thomson's 'plum-pudding' model of the atom

Negative electron

Positive matter

Ernest Rutherford

In 1909, Rutherford examined the results of Geiger and Marsden's experiment in which they had bombarded a very thin sheet of gold with positive alpha particles. The scientists recorded the pathway of the alpha particles through the gold. Rutherford found that while most alpha particles passed through the gold atoms undeflected, a small number of alpha particles were deflected a little and a tiny number of particles were deflected back towards the source. He concluded that the positive charge in the atom must be concentrated in a very small area of the atom. This area is the nucleus of the atom.

The Geiger–Marsden experiment helped Rutherford devise his 'nuclear' model of the atom

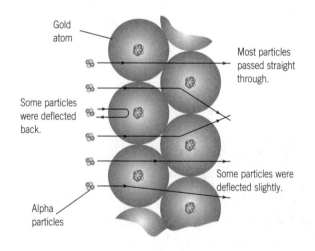

Gold atom

Most particles passed straight through.

Some particles were deflected back.

Some particles were deflected slightly.

Alpha particles

Neils Bohr

In 1913, Neils Bohr deduced that electrons must be found in certain areas in the atom otherwise they would spiral in towards the nucleus.

Atoms and the Periodic Table

Today, scientists consider the periodic table an important summary of the structure of atoms. The periodic table can be used to source the boiling point or density of elements. A detailed periodic table can be used to find the names, symbols, relative atomic masses and atomic number of any element.

In the periodic table, elements are arranged in order of increasing atomic number. It is called a periodic table because elements with similar properties occur at regular intervals or 'periodically'. The elements are placed in horizontal rows, called **periods**, and elements with similar properties appear in the same vertical column. These vertical columns are called **groups**. The elements in group 1 of the periodic table include lithium, sodium and potassium. All the elements in group 1 of the periodic table share similar properties: they are all metals except H

(hydrogen) and they all consist of atoms that have just one electron in their outer shell. When these metals react they form **ions**, which have a 1+ charge. Elements in the same period have the same number of shells of electrons.

All the isotopes of an element have the same number of electrons and protons. All the isotopes of an element appear in the same place on the periodic table.

✓ Maximise Your Marks

The modern periodic table is arranged in order of increasing atomic number. You must make this clear in your answers. Many students refer to increasing mass number, which is not correct.

The periodic table

Group	1	2											3	4	5	6	7	0
Period																		
1							H Hydrogen												He Helium
2	Li Lithium	Be Beryllium											B Boron	C Carbon	N Nitrogen	O Oxygen	F Fluorine	Ne Neon
3	Na Sodium	Mg Magnesium											Al Aluminium	Si Silicon	P Phosphorus	S Sulfur	Cl Chlorine	Ar Argon
4	K Potassium	Ca Calcium	Sc Scandium	Ti Titanium	V Vanadium	Cr Chromium	Mn Manganese	Fe Iron	Co Cobalt	Ni Nickel	Cu Copper	Zn Zinc	Ga Gallium	Ge Germanium	As Arsenic	Se Selenium	Br Bromine	Kr Krypton
5	Rb Rubidium	Sr Strontium	Y Yttrium	Zr Zicronium	Nb Niobium	Mo Molybdenum	Tc Technetium	Ru Ruthenium	Rh Rhodium	Pd Palladium	Ag Silver	Cd Cadmium	In Indium	Sn Tin	Sb Antimony	Te Tellurium	I Iodine	Xe Xenon
6	Cs Caesium	Ba Barium	La Lanthanum	Hf Hafnium	Ta Tantalum	W Tungsten	Re Rhenium	Os Osmium	Ir Iridium	Pt Platinum	Au Gold	Hg Mercury	Tl Thallium	Pb Lead	Bi Bismuth	Po Polonium	At Astatine	Rn Radon
7	Fr Francium	Ra Radium	Ac Actinium															

Key ☐ = non-metals ☐ = metals

? Test Yourself

1. Why do scientists have to re-evaluate existing models?

2. How are the elements arranged in the modern periodic table?

3. What are the horizontal rows and vertical columns in the periodic table called?

★ Stretch Yourself

1. John Dalton made a number of predictions about atoms. Today, which of his predictions is not thought to be correct? Explain your answer.

The Periodic Table

Early Ideas

As new elements were discovered, scientists struggled to find links between them.

In 1829, the German chemist Johann Wolfgang Dobereiner noticed that many elements could be put into groups of three, which he called **triads**. If these elements were placed in order of **atomic weight**, the middle element was about the average of the lighter and the heavier element.

He noticed a similar pattern when he compared the densities of the members of a triad. Unfortunately, this pattern only appeared to work some of the time.

History of the Periodic Table

In 1863, the English chemist **John Newlands** noticed that if the known elements were placed in order of their atomic weight, and then put into rows of seven, there were strong similarities between elements in the same vertical column.

This pattern became known as **Newlands' law of octaves**. It was useful for some of the elements, but unfortunately Newlands' pattern broke down when he tried to include the **transition elements**.

In 1869, the Russian chemist **Dimitri Mendeleev** produced his periodic table of elements. In his table, elements with similar properties occurred periodically and were placed in vertical columns called groups.

Like Newlands, Mendeleev arranged the elements in order of increasing atomic weight, but unlike Newlands he did not stick strictly to this order. He left gaps for elements that had yet to be discovered, such as germanium and gallium, and made detailed predictions about the physical and chemical **properties** these elements would have.

Eventually, when these elements were discovered and their properties analysed, scientists confirmed Mendeleev's predictions. His table went from being an interesting curiosity to a useful tool for understanding how a particular element would behave.

By leaving gaps and swapping the order of the elements, Mendeleev had actually arranged the elements in order of increasing atomic number (or the number of protons in the nucleus of an atom), even though protons themselves were not discovered until much later.

In fact, electrons, protons and neutrons were all discovered in the early 20th century.

Build Your Understanding

The Noble Gases

All the noble gases have similar properties (they all have single atoms), so they are in the same group in the periodic table. The noble gas group of elements include helium (He), neon (Ne), argon (Ar), krypton (Kr), xenon (Xe) and radon (Rn). The noble gases were only discovered in the 1890s when chemists noticed that the density of nitrogen made in reactions was slightly different to the density of nitrogen obtained directly from the air. The chemists thought the air might contain small amounts of other gases and so they devised experiments that eventually confirmed the presence of the very unreactive noble gases. They found that the air includes nitrogen, oxygen, neon and argon.

Transition Metals

Transition metals are found in the central block of the periodic table. The transition metals are much less reactive than group 1 metals.

Transition metals have high melting points so, with the exception of mercury, are solid at room temperature. They are hard and strong and make useful structural materials. They do not react with water or oxygen as vigorously as group 1 metals, although many will show signs of corrosion over long periods of time.

Transition metals are used in making a range of useful products such as aeroplanes and cars.

Uses of transition metals

Build Your Understanding

When transition metals form compounds the transition metal ions have variable charges. For example, in copper (II) oxide, CuO, the copper ions have a 2+ charge, while in copper (I) oxide, Cu_2O, the copper ions have a 1+ charge.

The roman numerals given in the name of the transition metal compound shows the charge on the transition metal ion. Transition metal compounds are coloured; group 1 and 2 metal compounds are white:

- Sodium is a group 1 metal.
- Sodium chloride is a white solid.

The compound sodium chloride

Transition metals and transition metal compounds are useful catalysts (chemicals that speed up chemical reactions). Iron is used in the Haber process (which produces ammonia) while nickel is used in the hydrogenation of unsaturated hydrocarbons.

Copper is a transition metal. Hydrated copper (II) sulfate is a blue solid. 'Hydrated' means that the compound contains water.

The compound hydrated copper (II) sulfate

❓ Test Yourself

1. What is the name given to Mendeleev's way of arranging the elements?
2. Why did Mendeleev not include the element germanium in his arrangement of the elements?

⭐ Stretch Yourself

1. What is the charge on the copper ion in the compound copper (II) oxide?

Chemical Reactions and Atoms

Symbols

Each element has its own unique symbol that is recognised all over the world.

Each symbol consists of one or two letters and is much easier to read and write than the full name.

In some cases the symbol for an element is simply the first letter of the element's name. This letter must be a capital letter: the element iodine is represented by the symbol I.

Occasionally, an element may take its symbol from its former Latin name. When this happens, the first letter is a capital and the second letter, if there is one, is lower case: the element mercury is represented by the symbol Hg. This comes from the Latin name for mercury, which was *hydrargyrum*, or **liquid silver**.

Several elements have names that start with the same letter. When this happens, the first letter of the element's name is used, together with another letter from the name. The first letter is a capital and the second letter is lower case: the element magnesium is represented by the symbol Mg.

✓ Maximise Your Marks

Remember to use the periodic table to check any symbols you are using. Don't forget that if a symbol has two letters, the first letter is a capital and the second is lower case.

Chemical Formulae

Compounds consist of two or more different types of atom that have been chemically combined.

A compound can be represented using a chemical **formula** that shows the type and ratio of the atoms that are joined together in the compound.

Ammonia has the chemical formula NH_3. This shows that in ammonia, nitrogen and hydrogen atoms are joined together in the ratio of one nitrogen atom to three hydrogen atoms.

You should take care when writing out the symbols for chemical compounds as some of them are very similar to elements. For example:

- The element carbon has the symbol C.
- The element oxygen has the symbol O.
- The element cobalt has the symbol Co.

The formula CO shows that a carbon atom and an oxygen atom have been chemically combined in a 1 : 1 ratio. This is the formula of the compound carbon monoxide.

The symbol Co represents the element cobalt. Notice how the second letter of the symbol is written in lower case. If it wasn't, it would be a completely different substance.

The formula CO_2 shows that carbon and oxygen atoms have been chemically combined in a 1 : 2 ratio. This is the formula of the compound carbon dioxide.

You need to be very careful when you write chemical symbols and formulae.

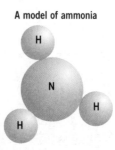
A model of ammonia

✓ Maximise Your Marks

Take care when writing formulae with subscript numbers. They will need to be perfect to get the mark awarded in the exam.

Chemical Reactions

Atoms can join together by:
- **Covalent bonding** – sharing pairs of electrons.
- **Ionic bonding** – giving and taking electrons.

Compounds formed from metals and non-metals consist of ions. These compounds are held together by strong ionic bonds. Compounds formed from non-metals often consist of molecules. The atoms are held together by strong **covalent** bonds.

Word and Symbol Equations

Symbol equations can be used to describe what happens during a chemical reaction.

When magnesium burns in air the magnesium metal reacts with the non-metal atoms in oxygen molecules to form the ionic compound magnesium oxide. This reaction can be shown in a word equation:

Magnesium + Oxygen → Magnesium Oxide

or by the symbol equation:

$2Mg + O_2 \rightarrow 2MgO$

Atoms are not created or destroyed during a chemical reaction: the atoms are just rearranged.

This means that the total mass of the **reactants** is the same as the total mass of the **products**.

Ionic Compounds

Ionic compounds are formed when a metal reacts with a non-metal. When metal atoms react they lose negatively charged electrons to become positively charged ions (or **cations**). When non-metal atoms react they gain negatively charged electrons to become negatively charged ions (or **anions**).

Ionic compounds

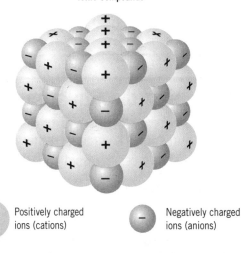

+	Positively charged ions (cations)	
–	Negatively charged ions (anions)	

Build Your Understanding

There is no overall charge on ionic compounds so you can use the charge on the ions to work out the formula of the ionic compound.

Metal Ions	Non-metal Ions
Sodium, Na^+	Bromide, Br^-
Potassium, K^+	Chloride, Cl^-

The compound sodium chloride contains sodium, Na^+, and chloride, Cl^-, ions. For every one sodium ion one chloride ion is required. The overall formula for the compound is NaCl.

? Test Yourself

1. How can atoms join together?
2. Give the name of the elements with the symbols Na and Cr.
3. A water molecule has the formula H_2O. Explain what this formula tells us.
4. Sodium nitrate has the formula $NaNO_3$. Explain what this formula tells us.

★ Stretch Yourself

1. Give the formula for the following compounds:
   a) Potassium chloride.
   b) Sodium bromide.

Balancing Equations

Conservation of Mass

Symbol equations show the type and ratio of the atoms involved in a reaction. The reactants are placed on the left-hand side of the equation. The products are placed on the right-hand side of the equation.

Overall, mass is **conserved** because atoms are never made or destroyed during chemical reactions. This means that there must always be the same number of each type of atom on both sides of the equation.

Balancing the Equation

Hydrogen burns in air to produce water vapour. This can be shown using a word equation:

Hydrogen + Oxygen → Water

The word equation is useful, but it doesn't give the ratio of hydrogen and oxygen molecules (small groups of atoms joined by covalent bonds – where atoms share pairs of electrons) involved. Balanced symbol equations show this extra information. First, replace the words with symbols.

Hydrogen and oxygen both exist as molecules:

$H_2 + O_2 \rightarrow H_2O$

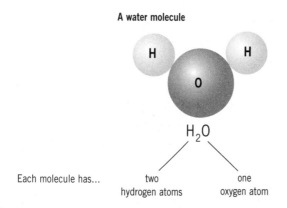

A water molecule

H_2O

Each molecule has... two hydrogen atoms   one oxygen atom

The formulae are all correct, but the equation does not balance because there are different numbers of atoms on each side of the equation. The formulae cannot be changed, but numbers can be added in front of the formulae to balance the equation.

The equation shows that there are two oxygen atoms on the left-hand side of the equation, but only one oxygen atom on the right-hand side.

A number 2 is therefore placed in front of the H_2O:

$H_2 + O_2 \rightarrow 2H_2O$

Now the oxygen atoms are balanced: there are the same number of oxygen atoms on both sides of the equation. However, the hydrogen atoms are no longer balanced. There are two hydrogen atoms on the left-hand side and four hydrogen atoms on the right-hand side. So in front of the H_2 a 2 is placed:

$2H_2 + O_2 \rightarrow 2H_2O$

The equation is now balanced.

When balancing equations, always check that the formulae you have written down are correct.

Some equations involve formulae that contain brackets, for example calcium hydroxide, $Ca(OH)_2$. This means that calcium hydroxide contains calcium, oxygen and hydrogen atoms in the ratio 1 : 2 : 2. These equations can be balanced normally.

✔ Maximise Your Marks

To get a top mark, you need to be able to balance equations. This skill just needs a little practice. Deal with each type of atom in turn until everything balances.

Remember to write any subscripts below the line: H_2O is correct while H^2O and H2O are wrong.

State Symbols

State symbols can be added to an equation to show extra information. They show what state the reactants and products are in.

The symbols are:
- (s) for solid
- (l) for liquid
- (g) for gases
- (aq) for aqueous.

Aqueous comes from the Latin *aqua* meaning water. Aqueous means dissolved in water.

Magnesium metal can be burned in air to produce magnesium oxide. Magnesium and magnesium oxide are both solids. The part of the air that reacts when things are burned is oxygen, which is a gas:

Magnesium + Oxygen → Magnesium Oxide

$$2Mg(s) + O_2(g) \rightarrow 2MgO(s)$$

Precipitation Reactions

A precipitate is a solid formed when two solutions react together.

Some **insoluble salts** can be made from the reaction between two solutions.

Barium sulfate is an insoluble salt. It can be made by the reaction between solutions of barium chloride and sodium sulfate:

Barium + Sodium → Barium + Sodium
Chloride   Sulfate   Sulfate   Chloride

$$BaCl_2(aq) + Na_2SO_4(aq) \rightarrow BaSO_4(s) + 2NaCl(aq)$$

Build Your Understanding

The insoluble salt barium sulfate can be filtered off, washed and dried.

Overall, the two original salts, barium chloride and sodium sulfate, have swapped partners. This can be described as a double decomposition reaction. Barium chloride solution can be used to test whether a solution contains sulfate ions. If sulfate ions are present, a white precipitate of barium sulfate will be seen.

The chloride ions and sodium ions are spectator ions. They are present, but they are not involved in the reaction. The ionic equation for the reaction is:

$$Ba^{2+}(aq) + SO_4^{2-}(aq) \rightarrow BaSO_4(s)$$

Precipitation reactions are very fast. When the reactant solutions are mixed, the reacting ions collide together very quickly and react together to form the insoluble solid.

Barium sulfate is used in medicine as a barium meal. The patient is given the insoluble salt and then X-rayed. The barium sulfate is opaque to X-rays so doctors can detect digestive problems without having to carry out an operation. Although barium salts are toxic, barium sulfate is so insoluble that very little dissolves and passes into the bloodstream of the patient.

An X-ray taken following a barium meal

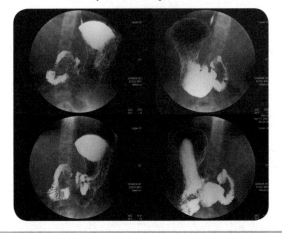

? Test Yourself

1. Why must there be the same number of each type of atom on both sides of an equation?
2. Balance the equation $Na + Cl_2 \rightarrow NaCl$.
3. Balance the equation $H_2 + Cl_2 \rightarrow HCl$.

★ Stretch Yourself

1. Explain why precipitation reactions happen very quickly.

4. What does the state symbol (l) indicate?

Ionic and Covalent Bonding

Types of Bonding

Compounds are made when atoms of two or more elements are chemically combined.

Ionic bonding involves the transfer of electrons in the outermost shell of atoms. This forms **ions** with opposite charges, which then attract each other. Ions are atoms or groups of atoms with a charge.

Covalent bonding involves the sharing of pairs electrons. The attraction between shared pairs of electrons holds the atoms together.

Ionic Bonding

Atoms react to get a full outer shell of electrons. Ionic bonding involves the transfer of electrons from one atom to another.

Metal atoms in groups 1 and 2, such as sodium or calcium, lose electrons to get a full outer shell of electrons. Overall, they become positively charged (electrons have a negative charge). Non-metal atoms in groups 6 and 7, such as oxygen or chlorine, gain electrons to get a full outer shell. They become negatively charged.

An ion is an atom, or a group of atoms, with a charge. An atom, or group of atoms, becomes an ion by gaining or losing electrons. The positive and negative ions formed have the same electronic structure as a noble gas atom.

Build Your Understanding

Ionic Compounds

Sodium reacts with chlorine to make sodium chloride:

Sodium + Chlorine → Sodium Chloride

- Each sodium atom transfers one electron from its outer shell to a chlorine atom.
- The sodium atom has lost a negatively charged electron, so it now has a 1+ charge and is called a sodium ion.
- The chlorine atom has gained an electron so it has a 1− charge and is now a chloride ion.
- Both the sodium ion and chlorine ion have a full outer shell.
- The attraction between these two oppositely charged ions is called an ionic bond, which holds the compound together.

In dot and cross diagrams, the electrons drawn as dots, and the electrons drawn as crosses are identical. They are drawn like this so it easier to see what happens when the electrons move.

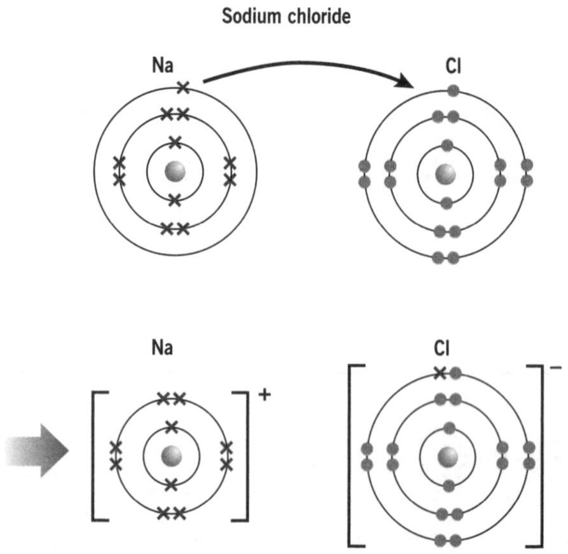

Sodium chloride

Na   Cl

Na   Cl

Atoms and Materials

Build Your Understanding (cont.)

Magnesium Oxide

Magnesium reacts with oxygen to make magnesium oxide:

Magnesium + Oxygen → Magnesium Oxide

- The magnesium atom transfers two electrons from its outer shell to the oxygen atom.
- The magnesium atom has lost two electrons so has a 2+ charge. It is now a magnesium ion.
- The oxygen atom has gained two electrons so has a 2– charge. It is now an oxide ion.
- Both the magnesium and oxygen atoms have a full outer shell.
- The attraction between these two oppositely charged ions is called an ionic bond, which holds the compound together.

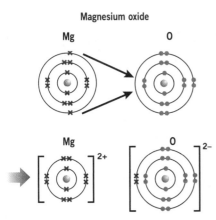

Magnesium oxide

Magnesium oxide has a higher melting point than sodium chloride because magnesium oxide contains smaller ions that have higher charges, so the attraction between these ions is stronger.

Make sure you can apply these ideas to other examples.

Covalent Bonding

Covalent bonding occurs between atoms of non-metal elements. The atoms share pairs of electrons so that all the atoms gain a full outer shell of electrons.

There is an **electrostatic attraction** between the nuclei of the atoms and the bonding electrons.

Hydrogen, H_2

Both the hydrogen atoms have just one electron. Both atoms can get a full outer shell by sharing their electrons to form a single covalent bond.

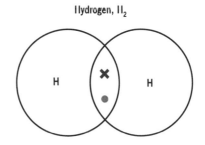

Hydrogen, H_2

Oxygen, O_2

Both oxygen atoms have six outer electrons so they share two pairs of electrons to form a double covalent bond.

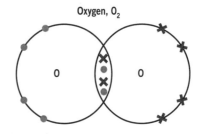

Oxygen, O_2

Make sure you can apply these ideas to other examples.

❓ Test Yourself

1. What are ions?
2. What happens to electrons during the formation of ionic bonds?
3. What holds the atoms together in covalent molecules?
4. What charge does a sodium ion have?

⭐ Stretch Yourself

1. Name and describe the type of bonding that you would expect to find in these substances:
   a) Oxygen, O_2.
   b) Sodium chloride, NaCl.

Ionic and Covalent Structures

Ionic Bonding

Ionic bonding occurs between metal and non-metal atoms. It involves the transfer of electrons and the formation of ions. Sodium chloride and magnesium oxide are examples of ionic compounds.

Build Your Understanding

Ionic compounds are held together by the strong forces of attraction between oppositely charged ions (electrostatic attraction).

Ionic compounds have a regular structure.

The strong forces of attraction between oppositely charged ions work in all directions and this means that ionic compounds have high melting and boiling points.

When dissolved in water, ionic compounds form solutions in which the ions can move so these solutions can conduct electricity.

Similarly, if ionic compounds are heated up so that they melt, the ions can move. Molten ionic compounds can also conduct electricity.

Simple Covalent Structures

Covalent bonding occurs between non-metal atoms. It involves the sharing of pairs of electrons. Examples of simple covalent structures include chlorine and oxygen.

These molecules are formed from small numbers of atoms. The low boiling points of simple molecules is the result of the weak forces of attraction between molecules.

Oxygen is an example of a simple covalent structure

Properties of Simple Covalent Structures

In simple covalent structures there are very strong covalent bonds between the atoms in each molecule, but very weak forces of attraction between these molecules.

This means that molecular compounds have low melting and boiling points; most are gases or liquids at room temperature. Simple molecular compounds do not conduct electricity because, unlike ions, the molecules do not have an overall electrical charge. They tend to be **insoluble** in water (although they may dissolve in other solvents).

💡 Boost Your Memory

To help you learn the facts you need for the exam, make a table to compare the features of ionic and covalent structures.

Giant Covalent Structures

Examples of giant covalent (macromolecular) structures include diamond, graphite and silicon dioxide.

These structures are formed from a large number of atoms.

Properties of Giant Covalent Structures

The atoms in giant covalent structures are held together by **strong covalent bonds**. This means that these substances have high melting and boiling points. They are solids at room temperature. Like simple covalent molecules, giant covalent substances do not conduct electricity (except graphite) as they do not contain ions. They are also insoluble in water.

Diamond

Diamond is an example of a giant covalent substance. Like other gemstones, it is prized for its rarity and its pleasing appearance: it is lustrous, colourless and transparent. Diamond is also very hard.

High quality diamonds are used to make jewellery. Other diamonds are used in industry in a variety of applications.

The hardness and high melting point of diamond makes it very suitable for cutting tools.

The special properties of diamond are the result of its **structure**.

Graphite

Allotropes are different forms of the same element in the same physical state. Diamond, graphite and fullerene are all allotropes of carbon.

Graphite is lustrous, black and opaque. It is used to make pencil 'leads' because the layers slide over each other easily, so when a pencil is rubbed on paper a black mark is left.

Graphite is also used in lubricants because it is slippery and allows surfaces to pass over each other more easily.

Diamond Bonding

Each carbon atom is bonded to four other carbon atoms by strong covalent bonds. It takes a lot of energy to break these strong bonds.

Diamond is a very poor electrical conductor because it does not have any free electrons.

Silicon dioxide, SiO_2, is found in the mineral quartz. It is very hard, which makes it resistant to weathering and it is the main constituent in sandstone. It is sometimes called silica.

Silicon dioxide has a similar structure to diamond and each silicon atom is attached to oxygen atoms by covalent bonds. Silicon dioxide does not conduct electricity because all the electrons are held in strong covalent bonds and cannot move.

Graphite Bonding

In graphite, each carbon atom forms strong covalent bonds with three other carbon atoms in the same layer. However, the bonding between layers is much weaker so the layers can pass over each other quite easily, which is why graphite is soft and feels greasy.

If a potential difference is applied across graphite, the electrons in the weak bonds between layers move and so conduct electricity.

Carbon in the form of graphite is the only non-metal element that conducts electricity. It also has a very high melting point because a lot of energy is required to break these strong covalent bonds. These properties make it a suitable material from which to make electrodes.

? Test Yourself

1 What type of structure is:
   a) magnesium oxide?
   b) graphite?
   c) diamond?

★ Stretch Yourself

1 Sodium chloride is an ionic compound. It does not conduct electricity when solid, but does when it is dissolved in water. Explain these observations in terms of the particles involved.

Group 7

The Halogens

The elements in group 7 are known as the **halogens**. The atoms of group 7 elements all have seven electrons in their outermost shell. When halogen atoms react they gain an electron to form **halide ions**.

Group 7 elements have similar properties because they all have similar electron structures. Halogens react with metal atoms to form ionic compounds, for example chlorine reacts with potassium to form potassium chloride. Group 7 atoms form molecules in which two atoms are joined together. These are called **diatomic** molecules.

The halogen family includes fluorine, chlorine, bromine and iodine. Halogens have coloured vapours. The colour gets darker further down the group.

Build Your Understanding

Down the group, the melting points and boiling points of the halogens increase, so fluorine and chlorine are gases at room temperature while bromine is a liquid and iodine is a solid.

Halogens react with hydrogen to form hydrogen halides, for example chlorine reacts with hydrogen to form hydrogen chloride. Hydrogen halides dissolve in water to form acidic solutions.

The atoms get larger and have more electrons further down the group. This means that the strength of the attraction between molecules increases.

As the forces of attraction between molecules get stronger down the group, it takes more energy to overcome these forces so the halogens will melt and boil at higher temperatures.

Physical Properties and Uses

Fluorine is a very poisonous gas that should only be used in a fume cupboard. Fluorine is a diatomic molecule with the formula F_2. The gas has a pale yellow colour.

Sodium fluoride is added to toothpastes and to some water supplies to help prevent tooth decay. Scientists carried out large studies to prove that adding fluoride compounds was effective at protecting teeth, but some people are concerned over the lack of choice those living in affected areas now have.

Chlorine is a poisonous gas that should only be used in a fume cupboard. Chlorine is a diatomic molecule with the formula Cl_2. The gas has a pale green colour.

Chlorine kills bacteria and is used in water purification. It is also used to make plastics and pesticides and in bleaching.

In the past, chlorine and iodine were extracted from compounds found in seawater. However, it is no longer economically worthwhile to extract iodine in this way.

Bromine is a poisonous, dense liquid. It has a brown colour.

Iodine exists as a black crystalline solid. Solid iodine is brittle, crumbly and is a poor electrical and thermal conductor. Iodine forms a purple vapour when warmed.

Iodine solution can be used as an antiseptic to sterilise wounds because it kills bacteria and can be used to test for the presence of starch. When iodine solution is placed on a material that contains starch it turns blue/black.

Astatine is found just below iodine in the periodic table. We can use the physical properties of the other halogens to predict the properties of astatine. It will be a dark coloured solid at room temperature.

Why Chlorine Reacts More Vigorously than Bromine

Reactivity decreases down the group: as an atom reacts to form an ion, the new electron is being placed into a shell further away from the nucleus. So, down the group, it is harder for atoms to gain an electron. There are also more shells of electrons shielding the new electron from the nucleus. This also makes it harder for atoms to gain a new electron further down the group. This pattern is clearly shown by the reaction between the halogens and iron wool to form iron halides.

Halogen Used	Observations
Chlorine	The iron glows very brightly. A brown smoke is given off and a brown solid is formed.
Bromine	The iron glows. Brown smoke is given off and a brown solid is formed.

💡 Boost Your Memory

Chlorine is more reactive than bromine and iodine.

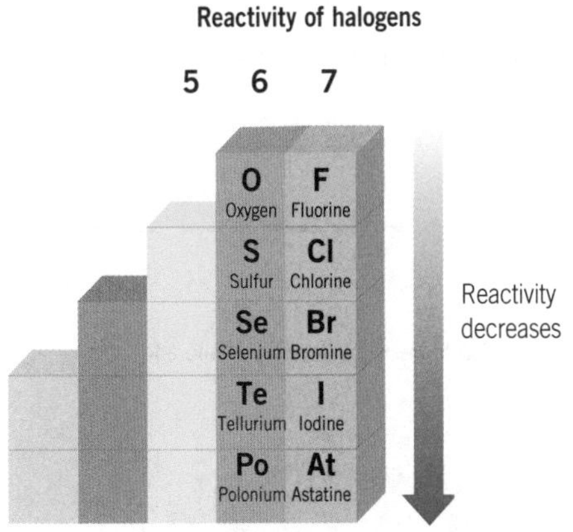

Reactivity of halogens

Displacement Reactions Involving Halogens

Reactivity decreases down group 7. The most reactive halogen is fluorine, followed by chlorine, then bromine, then iodine.

A more reactive halogen will **displace** (that is, take the place of) a less reactive halogen from an aqueous solution of its salt. So, chlorine could displace bromine and iodine.

However, while bromine could displace iodine it could not displace chlorine.

Chlorine will displace iodine from a solution of potassium iodide:

Chlorine + Potassium Iodide → Iodine + Potassium Chloride

$$Cl_2 + 2KI \rightarrow I_2 + 2KCl$$

Build Your Understanding

Chlorine reacts with potassium to make potassium chloride:

Potassium + Chloride → Potassium Chloride

$$2K + Cl_2 \rightarrow 2KCl$$

When they react, a halogen atom gains an electron to form an ion with a 1− charge:

$$Cl + e^- \rightarrow Cl^-$$

A reduction reaction has taken place. The halogen atom has gained an electron so it is reduced.

❓ Test Yourself

1. What is the name given to group 7 of the periodic table?
2. How is a halide ion formed?
3. What type of compound is formed when a metal reacts with a halogen?

⭐ Stretch Yourself

1. Chlorine gas is passed through an aqueous solution of potassium iodide. Write word and symbol equations to sum up this reaction.

New Materials

Smart Materials

Many scientists are involved in making new ('smart') materials, which can have very special properties. Smart materials have one or more property that can be dramatically, and reversibly, altered by changes in the environment.

Scientists are working to find more applications for smart materials and to discover new materials.

A whole variety of smart materials already exist including shape-memory alloys, thermochromic materials and photochromic materials.

Build Your Understanding

Photochromic materials change colour when exposed to bright light. They are widely used to make lenses for glasses. The lenses adapt to light conditions: when it is bright, the lenses get darker.

Photochromic glasses

Hydrogels

Hydrogels are a new type of polymer. They are able to absorb water and swell up as the result of changes in pH or changes in temperature. Hydrogels are being used to make special wound dressings. They help to:

- Stop fluid loss from the wound.
- Absorb bacteria and odour molecules.
- Cool and cushion the wound.
- Reduce the number of times the wound has to be disturbed.

The hydrogel is transparent, so medical staff can monitor the wound without having to remove the dressing.

Buckminster Fullerene, C_{60}

The element carbon exists in three forms or allotropes:
- graphite
- diamond
- fullerenes.

Fullerenes are structures made when carbon atoms join together to form tubes, balls or cages, which are held together by strong covalent bonds. The most symmetrical and most stable example is buckminster fullerene. This is a new material scientists have discovered, which consists of 60 carbon atoms joined together in a series of hexagons and pentagons, much like a leather football.

Structure of buckminster fullerene

✓ Maximise Your Marks

Learn the formula for buckminster fullerene: C_{60}.

Nanoparticles

Nanoscience is the study of extremely small pieces of material called **nanoparticles**. Scientists are currently researching the properties of new nanoparticles.

These are substances that contain just a few hundred atoms and vary in size from 1 nm (nanometres) to 100 nm (human hair has a width of about 100 000 nm). Nanoparticles occur in nature, for example in sea spray. They can also be made accidentally, for example when fuels are burned.

Nanomaterials have unique properties because of the very precise way in which the atoms are arranged. Scientists have found that many materials behave differently on such a small scale.

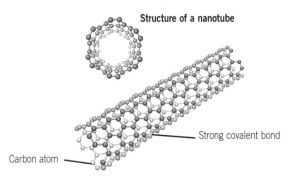

Structure of a nanotube

Strong covalent bond

Carbon atom

Lightweight Materials

Scientists are using nanoparticles to develop lightweight materials. These materials are incredibly hard and strong because of the precise way that the atoms are arranged. One day these materials could be used to build planes.

✓ Maximise Your Marks

In nanomaterials, the atoms themselves are not smaller. When you answer exam questions, make sure you do not infer that the atoms have changed size.

Other Uses of Nanoparticles

Nanoparticles have a very high surface area to volume ratio. Scientists hope that this will allow them to use nanoparticles in exciting ways such as:
- In new computers.
- In sunscreens and deodorants.
- In drug delivery systems.
- As better catalysts.

Catalysts are substances that speed up the rate of a chemical reaction, but are not themselves used up. Reactions take place at the surface of the catalyst. The larger the surface area of the catalyst, the more changes can take place at once and the better the catalyst performs.

Scientists are also keen to explore the use of nanoparticles as sensors to detect biological or chemical agents at very low levels. They may also be used to make battery electrodes for electric vehicles or solar cells.

Nanoscale silver particles have antibacterial, antiviral and antifungal properties. These tiny pieces of silver are incorporated into materials to make clothes and medical dressings stay fresh for longer.

There has recently been a great deal of media interest in the development and applications of new nanoparticles. Some scientists are concerned that certain nanoparticles could be dangerous to people because their exceptionally small size may mean they are able to pass into the body in previously unimaginable ways, and could go on to cause health problems.

⚑ Boost Your Memory

Try producing a set of revision cards to learn the important ideas in this topic.

❓ Test Yourself

1. Why are smart materials special?
2. How big are nanoparticles?
3. What is the formula of buckminster fullerene?
4. Where are nanoparticles found in nature?

⭐ Stretch Yourself

1. What is special about a photochromic material?

Synthesis

Making New Chemicals

The manufacture of useful chemicals involves many stages. **Raw materials** need to be selected and prepared, and then the new chemicals have to be made in a process known as **synthesis**.

Next, the useful products have to be *separated* from **by-products** and waste, each of which must also be dealt with. Finally, the **purity** of the product must be checked.

Build Your Understanding

Some chemicals are made in batch processes which are used to make relatively small amounts of special chemicals such as medicines. The chemicals are made when they are needed, rather than all the time.

Continuous processes are used to make chemicals that are needed in large amounts, such as sulfuric acid or ammonia. These chemicals are made all the time. Raw materials are continuously added and the new products are removed.

Some chemicals, such as ammonia, sulfuric acid, sodium hydroxide and phosphoric acid, are made in bulk (on a large scale). Other chemicals, such as medicines, food additives and fragrances, are described as being made on a fine scale (a small scale).

Governments regulate how chemicals are made, stored and transported to protect people and the environment from accidental damage.

Medicines from Plants

Scientists can extract chemicals from plants or produce them synthetically. Chemicals can be extracted from plants by:
- Crushing up the plant material.
- Adding a suitable solvent and then heating the mixture so that the useful chemicals dissolve in the solvent.
- Using separation techniques, such as **chromatography**, to separate mixtures of compounds. Chromatography separates mixtures according to differences in solubility of the components.

Scientists can use the melting point and boiling point of a compound to establish its purity. **Thin layer** chromatography can also be used.

Plant materials can be used to make very useful medicines. Digitalin medicines are extracted from foxglove plants and are used to treat heart conditions.

Morphine is made from opium poppies and is used for pain relief.

Scientists have discovered that corn starch can be used to make biodegradable plastics. These plastics are useful as they break down more easily in the environment.

Digitalin can be extracted from foxglove plants

Boost Your Memory

Make a flow diagram to show the stages involved in extracting chemicals from plants.

Making and Developing New Medicines

New medicines are often very expensive to buy because of the high costs of developing and making the drugs.

The factors that affect the price of a medicine include:
- Labour and energy costs. The production of new medicines is often very labour intensive as little automation is possible, at least initially.
- The cost of the raw materials required, which may be very rare or expensive.

- The time required for researching and developing new drugs. These processes can take many years.
- Testing of the new medicine. It must pass all the testing stages and human trials required by law for it to gain a licence to be sold. This takes a lot of time and money.
- Marketing of the medicine. Companies have to let the medical profession know the benefits of the new medicine.

Build Your Understanding

Scientists developing new drugs need to be aware of economic considerations. The more research and development involved, the more expensive the new medicine will be. Scientists need to work out if there is sufficient demand for the new medicine for it to pay back the considerable investment needed to produce it. New drugs only have a patent for a certain length of time. Companies manufacturing the medicine pay money to the people who hold the patent and who did the initial research and development for the drug. If the time limit for the patent is set too low, the patent will have run out before the initial costs have been paid back.

Green Chemistry

The long-term sustainability of a chemical process depends on:
- Whether or not the raw materials are renewable.
- The atom economy of the reactions involved.
- The amount and nature of waste produced.
- The amount and nature of by-products produced.
- The energy requirements.
- The impact on the environment.

- The health and safety risks.
- The economic and social benefits of the products made by the reaction.

✓ Maximise Your Marks

Addition reactions (when two chemicals are joined together), such as the reaction between ethene and steam to produce ethanol, will have an atom economy of 100 per cent.

❓ Test Yourself

1. Why are the labour costs for new medicines often very high?
2. Why do new drugs have to be marketed?
3. Name a separation technique that separates mixtures because the components have different solubilities.

⭐ Stretch Yourself

1. Describe what happens during a continuous process.

Practice Questions

Complete these exam-style questions to test your understanding. Check your answers on page 120. You may wish to answer these questions on a separate piece of paper.

1 Use the periodic table on page 7 to answer the questions below.

   a) Complete the diagram to show
      the electron structure of magnesium.   (1)

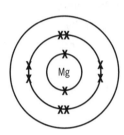

   b) Explain why magnesium is in group 2 of the periodic table.   (1)

   c) The chemical equation for the reaction between magnesium and bromine is given below.

       $Mg + Br_2 \rightarrow MgBr_2$

       Describe what happens in this reaction. Name the substances and give the number of
       atoms involved.   (2)

   d) Explain why magnesium bromide conducts electricity when molten but not when solid.   (2)

2 Ammonia has the formula NH_3.

   a) Describe the bonding in an ammonia molecule.   (1)

   b) Explain why ammonia is a gas at room temperature.   (1)

3 The diagrams below show two atoms of the element oxygen. Atoms of oxygen contain three different types of particle. The particles in the atoms have been represented by the symbols **x**, ⚪ and ⚫.

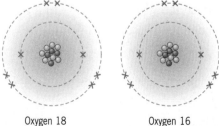

         Oxygen 18   Oxygen 16

a) What type of particles do the crosses on the diagrams represent? (1)

....

b) i) What is the centre of an atom called? (1)

....

ii) Which types of particles do the dots in the centre of the oxygen atoms represent? (1)

....

c) The table below shows some information about the two oxygen atoms in the diagrams shown above. Complete the table to show the number of protons, neutrons and the electron structure of each of the oxygen atoms. (5)

	Number of Protons	Number of Neutrons	Electron Structure
$^{16}_{8}O$		8	
$^{18}_{8}O$			

4 Here is diagram of a methane, CH_4, molecule.

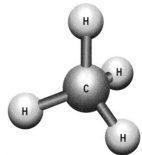

a) What type of bonding is in methane molecules? (1)

....

b) Explain why methane is a gas at room temperature. (1)

....

5 Nanoparticles are substances that are just a few hundred atoms in size. They could be very useful but some people are concerned about the development and use of these materials. Use your knowledge and understanding of the topic to outline the advantages and disadvantages of developing and using nanoparticles.

You should make sure your answers are written using good spelling, punctuation and grammar. (6)

....

....

....

....

Atoms and Materials

How well did you do?

| 0–8 | Try again | 9–15 | Getting there | 16–20 | Good work | 21–24 | Excellent! |

Evolution of the Atmosphere

The Atmosphere Today

Today, the Earth's **atmosphere** is composed of:
- About 78 per cent **nitrogen**.
- About 21 per cent **oxygen**.
- Small amounts of other gases, such as carbon dioxide, water vapour and **noble gases**, for example argon and neon.

Build Your Understanding

Small changes in today's atmosphere can be caused by volcanic activity or by human activities, such as deforestation or farming. Throughout the history of the Earth, the composition of the atmosphere has changed.

Formation of the Atmosphere

The First Billion Years
- During the first billion years of the Earth's life, there was a great deal of volcanic activity which belched out carbon dioxide (CO_2), steam or water vapour (H_2O), ammonia (NH_3) and methane (CH_4).
- The atmosphere consisted mainly of carbon dioxide and there was very little oxygen. The atmosphere was very similar to that of the planets Mars and Venus today.
- Steam **condensed** to form the early oceans.

The Next Two Billion Years
- During the two billion years that followed, plants and **algae** evolved and began to cover the surface of the Earth.
- The plants grew very well in the carbon dioxide rich atmosphere. They steadily removed carbon dioxide and produced oxygen (O_2).
- Most of the carbon dioxide in the early atmosphere dissolved into the oceans. The carbon gradually became locked up in the shells and skeletons of marine organisms and, when the organisms died, as **carbonate** minerals. Some of the carbon from the early atmosphere is also stored in fossil fuels.
- The ammonia in the early atmosphere reacted with oxygen to release nitrogen. Living organisms, such as denitrifying bacteria, also produced nitrogen.

- As the amount of oxygen increased, an **ozone layer** (O_3) developed. This layer filtered out harmful **ultraviolet (UV) radiation** from the Sun, enabling new, more complex life forms to develop.

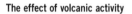

The effect of volcanic activity

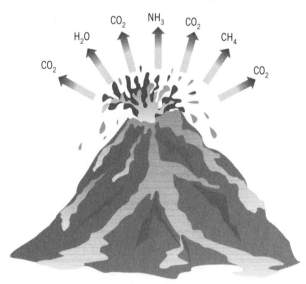

✔ Maximise Your Marks

If you are asked to suggest why scientists have slightly different ideas about the evolution of the atmosphere, remember that it is difficult for scientists to be completely precise about all the details because different sources of information suggest slightly different things may have happened.

Chemical Theories for the Origins of Life

People used to believe that life started when living things were generated spontaneously from non-living materials. Today, many scientists believe that life started because of chemical reactions taking place, for example between hydrocarbons, ammonia and lightning. The Earth's early environment would have provided the necessary conditions and raw materials (or primordial soup) for life to develop. There are many theories about how life first evolved.

In the **Miller–Urey experiment**, water, methane, ammonia and hydrogen were placed into sterile flasks and exposed to electrical sparks. Miller and Urey found that the reaction produced amino acids.

Carbon Dioxide and Fossil Fuels

The level of carbon dioxide in our atmosphere has increased as we have burned more fossil fuels.

These fossil fuels had stored carbon from the Earth's early atmosphere for hundreds of millions of years.

There is a mismatch, however, between the amount of carbon dioxide released into the atmosphere by the burning of fossil fuels and the actual increase in the amount of carbon dioxide in the atmosphere. A great deal of the carbon dioxide appears to be missing.

Build Your Understanding

Carbon dioxide is removed from the atmosphere by:
- Photosynthesis by plants on land.
- Photosynthesis by phytoplankton in the oceans.
- Dissolving of carbon dioxide from the atmosphere into the oceans.

The carbon dioxide reacts with seawater to produce:
- Insoluble (does not dissolve) carbonate salts, which are deposited as sediment.
- Soluble (does dissolve) calcium and magnesium hydrogen carbonate salts.

Much of the carbon dioxide is locked up in sediment for long periods of time. Some of this carbon dioxide is later returned to the atmosphere when the sediment is forced underground by geological activity and then released when volcanoes erupt.

However, not all of the carbon dioxide released by the burning of fossil fuels is removed in these ways. Many people are concerned about rising levels of carbon dioxide in the Earth's atmosphere and the possible link between these increased levels and global warming.

❓ Test Yourself

1. Approximately how much of today's atmosphere is made up of oxygen?
2. What was the main gas in the Earth's early atmosphere?
3. How did the evolution of plants affect the Earth's atmosphere?

⭐ Stretch Yourself

1. Describe the experiment carried out by Miller and Urey and explain why the result of the experiment was so significant.

Earth and Pollution

Noble Gases and the Fractional Distillation of Air

The Level of Oxygen Present in Air

Air is a mixture of different gases including nitrogen, oxygen and about 1 per cent of other gases including argon, neon, water vapour and carbon dioxide. All of these molecules and atoms are very small. There are large spaces between them:

- Nitrogen is an element. Nitrogen exists as molecules, with the formula N_2.
- Oxygen is also an element and exists as molecules, with the formula O_2.
- Argon and neon are both elements that exist as single atoms. They are represented by the symbols Ar and Ne.

- Water vapour is a molecular compound, with the formula H_2O.
- Carbon dioxide is also a molecular compound, with the formula CO_2.

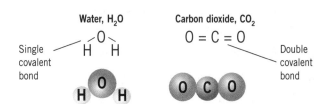

Water, H_2O

Single covalent bond

Carbon dioxide, CO_2

$O = C = O$

Double covalent bond

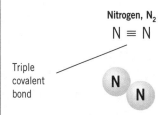

Nitrogen, N_2

$N \equiv N$

Triple covalent bond

Oxygen, O_2

$O = O$

Argon

Ar

Neon

Ne

Build Your Understanding

Combustion, or burning, reactions need oxygen.

Fuels will burn better in pure oxygen than they do in air. Oxygen can be mixed with the fuel acetylene in welding torches.

If an upturned beaker is placed over a candle, the oxygen in the air is used up as the candle burns and the water level inside the beaker rises.

When the candle uses up all the oxygen in the air the candle goes out. The water level moves about a fifth of the way up the beaker. This shows that about a fifth, or around 20 per cent, of the air is oxygen.

Combustion

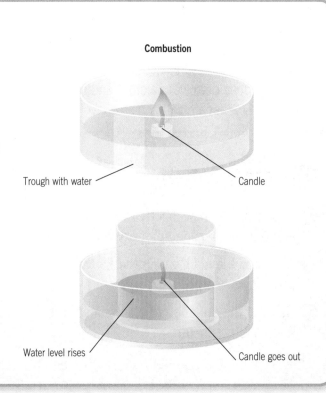

Trough with water

Candle

Water level rises

Candle goes out

Fractional Distillation of Liquid Air

Fractional distillation separates mixtures into different fractions, or parts, because the fractions have different boiling points. Both oxygen and nitrogen can be extracted from the air by fractional distillation.

Build Your Understanding

In fractional distillation, the air is filtered to remove dust and other impurities.

Next, the air is cooled until it reaches −200 °C. The gases condense to form liquids.

Carbon dioxide and water are removed, leaving a mixture of liquid nitrogen and oxygen. Oxygen turns from a liquid to a gas at −183 °C, while nitrogen turns from a liquid to a gas at −196 °C. The liquefied air mixture is fed into a fractionating column.

A final step is required to remove traces of argon and neon from the oxygen. These two gases have such similar boiling points that a second fractional distillation step is required to separate them. The gases separated from air are useful raw materials.

Separating oxygen and nitrogen

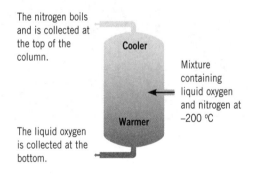

The nitrogen boils and is collected at the top of the column.

Cooler

Mixture containing liquid oxygen and nitrogen at −200 °C

Warmer

The liquid oxygen is collected at the bottom.

Noble Gases

The elements of group 0 are called the noble gases. Noble gases are very **unreactive**. They are sometimes described as being **inert** because they do not react. This is because they have a full, stable outer shell of electrons.

Notice that the noble gases have eight electrons in their outer shell, except for helium. Helium only has two electrons, but that still gives it a full outer shell.

Noble gases are useful to us precisely because they do not react. They are inert, have a **low density** and are **non-flammable**.

A model showing outer shell of electrons of the noble gases

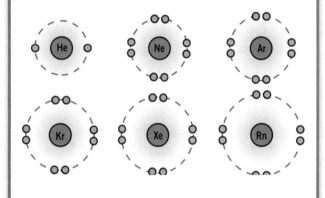

Uses of Noble Gases

- **Helium** is used in balloons and in airships because it is less dense than air.
- **Neon** is used in electrical discharge tubes in advertising signs.
- **Argon** is used in filament light bulbs. The hot filament is surrounded by argon. This stops the filament from burning away and breaking the bulb.

? Test Yourself

1. What is the formula of carbon dioxide?
2. Why can the components of liquid air be separated by fractional distillation?
3. How many electrons does a helium atom have?

★ Stretch Yourself

1. Outline the steps involved in the fractional distillation of air.
2. Why is it better to use helium than hydrogen in balloons?

Pollution of the Atmosphere

Earth and Pollution

Fossil Fuels

Coal is mainly carbon. Petrol, diesel, oil and natural gas are **hydrocarbons**.

When hydrocarbons are burned in a good supply of oxygen, water vapour and carbon dioxide are produced. The carbon and hydrogen atoms in the hydrocarbon fuels combine with the oxygen atoms. This is an example of an **oxidation** reaction:

Methane + Oxygen → Carbon Dioxide + Water Vapour

$$CH_4 + 2O_2 \rightarrow CO_2 + 2H_2O$$

Fuels will burn more quickly in pure oxygen than they will in air. The products made in these reactions can be pollutants that affect air quality.

Acid Rain

Fossil fuels, such as coal, oil and gas, often contain small amounts of sulfur. When these fuels are burned, the gas sulfur dioxide, SO_2, is produced. This gas can dissolve in rainwater to form **acid rain**. Acid rain can affect the environment by damaging buildings, as well as harming plant and aquatic life and corroding metals.

Fossil fuels are burned in power stations to produce electricity. So, using less electricity, by turning off lights when they are not in use and not leaving everyday appliances on standby mode, will help to reduce the amount of acid rain produced.

The damaging effect of acid rain on plant life

Build Your Understanding

Sulfur compounds can be removed directly from oil and gas before they are burned so that they do not produce sulfur dioxide as they burn. The sulfur that is removed is a valuable material that can be sold on.

It is more difficult to remove sulfur from coal. However, the sulfur dioxide produced by burning coal can be removed from the waste gases before they are released into the atmosphere. This process is carried out by scrubbers in power stations. The scrubbers react sulfur dioxide (from the waste gases) with calcium carbonate to produce gypsum and carbon dioxide. Sulfur dioxide can also be removed by oxidation and reaction with ammonia or by using seawater.

Carbon Monoxide

When fossil fuels containing carbon and hydrogen are burned, carbon dioxide and water vapour are produced. However, if a hydrocarbon fuel is burned in an insufficient supply of oxygen, carbon monoxide may also be produced. The gas carbon monoxide, or CO, can be a dangerous pollutant. It is colourless, odourless and very poisonous.

Incomplete combustion is undesirable because:
- It produces toxic carbon monoxide.
- Less heat is given off than when the fuel is burned fully.
- Soot is produced, which must then be cleaned. A sooty flame has a yellow colour.

Global dimming is caused by smoke particles that are released into the atmosphere. Scientists believe that these smoke particles reduce the amount of sunlight that reaches the Earth's surface and may even affect weather patterns.

✓ Maximise Your Marks

Practise writing balanced symbol equations for the incomplete combustion of fuels:

Methane + Oxygen → Carbon + Carbon + Water
Monoxide Vapour

$$3CH_4 + 4O_2 \rightarrow C + 2CO + 6H_2O$$

Catalytic Converters

In the UK, modern, petrol-fuelled cars are fitted with **catalytic converters** or 'cats'. The catalytic converter is part of a car's exhaust system and helps to reduce the amount of harmful gases that the car releases into the atmosphere. Catalytic converters work best at high temperatures and have a high surface area to increase the rate at which harmful gases are converted.

Build Your Understanding

Catalytic converters work in several ways:
- They help to convert carbon monoxide to carbon dioxide. The carbon monoxide is oxidised and the nitrogen monoxide is reduced:
$$2CO + 2NO \rightarrow N_2 + 2CO_2$$
- They help to convert nitrogen oxides to nitrogen and oxygen and they oxidise unburned hydrocarbons to carbon dioxide and water vapour.

Nitrogen oxide, NO, is produced at the high temperatures in the engine when nitrogen in the air is oxidised by the oxygen in the air. The nitrogen oxide reacts with oxygen to form nitrogen dioxide. Nitrogen oxide and nitrogen dioxide are referred to as nitrogen oxides or NOx. Nitrogen oxides cause acid rain and photochemical smog.

Atmospheric pollution caused by cars can also be reduced by:
- Having more efficient car engines.
- Encouraging people to make more use of public transport.
- Setting legal limits for the levels of pollutants in exhaust gases (these levels are checked during MOT tests).

? Test Yourself

1. What is the name of the gas produced when sulfur is burned?

2. How can you tell if a Bunsen burner is not completely burning the methane in its flame?

★ Stretch Yourself

1. Use a balanced symbol equation to show how a catalytic converter can help to remove carbon monoxide from the exhaust gases of a car.

The Greenhouse Effect and Ozone Depletion

Earth and Pollution

Carbon Dioxide and the Greenhouse Effect

The **greenhouse effect** is believed to be slowly heating up the Earth.

When fossil fuels are burned carbon dioxide is produced. Although some of this carbon dioxide is removed from the atmosphere by the reaction between carbon dioxide and seawater, the overall amount of carbon dioxide in the atmosphere has increased over the last 200 years.

Carbon dioxide traps the heat energy that has reached the Earth from the Sun.

Global warming may mean that the polar ice caps will eventually melt and this could cause massive flooding.

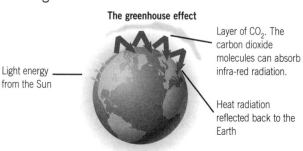

The greenhouse effect

Light energy from the Sun

Layer of CO_2. The carbon dioxide molecules can absorb infra-red radiation.

Heat radiation reflected back to the Earth

Reducing the Levels of Carbon Dioxide in the Atmosphere

In **carbon capture and storage**, the carbon dioxide produced by power stations is captured and then stored safely. The carbon dioxide can be stored in **porous** rocks. Carbon dioxide can also be converted into carbonate rocks and then stored.

Not all scientists believe that human activity is causing global warming. Other factors, such as solar cycles, may also be important. Scientists are also investigating ways to control the amount of carbon dioxide by:
- Adding iron to the oceans to encourage algae to **photosynthesise** and use up carbon dioxide.
- Converting carbon dioxide into useful hydrocarbons.

Build Your Understanding

Plants use carbon dioxide during photosynthesis so **deforestation** reduces the amount of carbon dioxide that can be removed from the air.

Organisms release carbon dioxide during respiration. Methane and water vapour are also greenhouse gases.

Ozone

Chlorofluorocarbons (CFCs) are organic molecules that contain carbon, chlorine and fluorine atoms. Chlorofluorocarbons:
- Are chemically inert and, therefore, non-toxic.
- Have low boiling points.
- Do not dissolve in water.

In the past, CFCs were widely used as aerosol propellants, as coolants in refrigerators and freezers and solvents used in dry-cleaning. Scientists believed the chemicals were safe to use. Then they found a link between CFCs and the depletion of the ozone layer.

Ozone (cont.)

Scientists discovered that the ozone layer was being damaged and they persuaded the rest of the world community that CFCs were responsible and should be replaced. They worked hard to find replacements.

Ozone, O_3, molecules consist of three oxygen atoms. The ozone layer is part of the stratosphere, which is in the Earth's atmosphere. Ozone in the stratosphere is important because it filters out harmful ultraviolet (UV) radiation and prevents it from reaching the lower part of the atmosphere. If more UV light reaches the lower part of the atmosphere, people could suffer medical problems including sunburn, skin damage and ageing, skin cancer and development of **cataracts**.

CFCs have been banned from use in almost all new products, but old fridges and freezers produced before the ban will still contain CFC molecules.

CFC molecules are stable because the carbon–halogen bonds are very strong and hard to break. However, CFC molecules do break down slowly in the Earth's stratosphere. UV light breaks down the strong covalent bonds in the CFC molecules. This process makes chlorine free radicals: chlorine atoms that are highly reactive because they have an unpaired electron. The chlorine free radicals are involved in the breakdown of ozone molecules, leading to the depletion of the ozone layer.

✓ Maximise Your Marks

Do not just state that UV causes cancer. You need to be clear that UV causes skin cancer.

Build Your Understanding

The steps involved in the breakdown of ozone molecules are shown below:

$$CFCl_3 \rightarrow Cl\cdot + \cdot CFCl_2$$

The chlorine free radical then reacts with an ozone molecule:

$$Cl\cdot + O_3 \rightarrow ClO\cdot + O_2$$
$$ClO\cdot + O \rightarrow Cl\cdot + O_2$$

Overall, ozone molecules are broken down:

$$O_3 + O \rightarrow 2O_2$$

Only a small number of chlorine free radicals need to be produced to have a dramatic effect because a chain reaction is set up. In this way, a single CFC molecule can destroy many ozone molecules. Ozone molecules absorb some types of UV radiation.

Alternatives to CFCs

Scientists have developed alternative compounds that can be used in place of CFCs, and do much less damage to the ozone layer.

Hydrochlorofluorocarbons, or HCFCs, have now widely replaced CFCs. HCFCs are non-toxic and non-flammable but they are very potent greenhouse gases.

❓ Test Yourself

1. How does adding iron to oceans reduce levels of carbon dioxide?

2. What do the initials CFC stand for?

3. How were CFCs used?

⭐ Stretch Yourself

1. Outline the steps involved in the breakdown of ozone molecules by chlorine free radicals.

Pollution of the Environment

Pollution

All substances are made from matter obtained from the Earth's crust, the sea or the atmosphere. It is vital that we protect the environment from harmful pollution. Some chemicals persist for long periods of time in the environment. Such chemicals can be carried over large distances and may even accumulate in human tissues through food consumption. Today, there are a large number of chemicals in the environment that may adversely affect the environment or human health, but scientists do not have enough data yet to be sure of all their effects.

Nitrate Fertilisers

Nitrogen, potassium and phosphorus are all needed for plants to grow. Nitrate fertilisers help plants to grow well and this means that farmers are able to produce more food.

Nitrate fertilisers can cause problems if they are washed into lakes or streams leading to eutrophication. This can cause environmental problems.

Nitrate fertilisers can also find their way into drinking water supplies. There have been health concerns over the levels of nitrates in water.

Problems with Oil Exploitation

Oil is an extremely important raw material. It is found in porous rocks in the Earth's crust. Sometimes, the crude oil has to be pumped up to the surface before it can be collected. It is often transported around the world in giant oil tankers. When accidents occasionally occur, crude oil can escape. The oil forms a slick that can devastate animal and plant life. For example, sea birds can die if their feathers become covered in oil. These oil slicks can do great damage to affected beaches; this can have serious consequences for local people, particularly in holiday areas. The detergents used to break up the oil slicks may also affect wildlife.

Political Considerations

In addition to the environmental problems associated with the exploitation of crude oil, there are also political problems that must be considered. Oil reserves are often found in politically sensitive areas. These countries may not want to sell great quantities of oil as a shortage in the world supply will naturally lead to an increase in the price of the oil. In other areas the situation can be even more unstable: battles in areas around oil fields can make it too dangerous for the oil to be extracted safely.

Problems with Bauxite Quarrying

Aluminium is extracted from its ore, bauxite. Unfortunately, this ore is often found in environmentally sensitive areas such as the Amazon rainforest. Bauxite is extracted from large opencast mines. Large numbers of trees must be cut down to clear space for the mines and new roads built to give access to them. In addition, litter and oil can pollute the area around the mines. However, aluminium can be recycled to reduce the need for bauxite quarrying.

Problems with Limestone Quarrying

The economic benefits of quarrying for limestone must be balanced against the environmental consequences of quarrying. Limestone has to be blasted from hillsides in huge quantities. This scars the landscape, causes noise pollution and dust, and affects local wildlife. Transporting limestone from the quarry can also cause problems, with heavy lorries causing noise, congestion and damaging local roads. Quarrying does, however, create new jobs and brings new money into an area.

✓ Maximise Your Marks

Make sure you can recall some of the advantages and disadvantages of limestone quarrying.

Plastics

Plastics are very useful materials:
- They are very stable and unreactive.
- Most plastics do not react with water, oxygen or other common chemicals.

Many plastics are non-biodegradable: they are not decomposed by microorganisms. Items made of plastic may have a greater environmental impact than their non-plastic alternatives.

Life cycle assessment (LCA) is used to assess the environmental impact an object has over its whole lifetime. This is sometimes referred to as 'from cradle to grave'. LCAs are an effective way of comparing several possible alternative products to see which one has the least impact on the environment. Recycling an object, if possible, will reduce its adverse effect on the environment.

Build Your Understanding

Plastics are made from substances obtained from crude oil.

Plastics can be disposed of by burning, but this solution may cause pollution problems. The common plastic PVC releases the gas hydrogen chloride when it is burned and can have harmful effects in the environment.

In response to these problems, scientists have developed new, biodegradable plastics that will eventually rot away. Some biodegradable plastics have been made from corn starch, while others can be disposed of by dissolving them in water. Scientists are also developing ways to recycle plastics.

Polyesters, which can be used to make fabrics and bottles, can be recycled to form fleece material to make new clothes. Recycling also means that our finite resources of crude oil will last longer.

To calculate the overall effect of an object on the environment, scientists measure the impact of:
- Extracting the raw materials.
- The manufacturing process.
- Any packaging used.
- How the product is transported.
- How it is used.
- What happens to the object when it is no longer useful.

? Test Yourself

1. Name the main ore of aluminium.
2. How is crude oil transported around the world?
3. How can limestone extraction affect the landscape?

★ Stretch Yourself

1. Suggest what the plastic polyester can be recycled into.

Evidence for Plate Tectonics

The Structure of the Earth

Scientists believe that the Earth has a layered structure:

- The outer layer, called the **crust**, is very thin and has a low **density**.
- The next layer down is called the **mantle**. This layer extends almost halfway to the centre of the Earth. The rock in the mantle is mainly solid, but small amounts must be liquid as the mantle flows very slowly.
- At the centre of the Earth is the **core**. The core consists of two parts: the outer core is liquid; the inner core, which is under enormous pressure, is solid.

The layered structure of the Earth

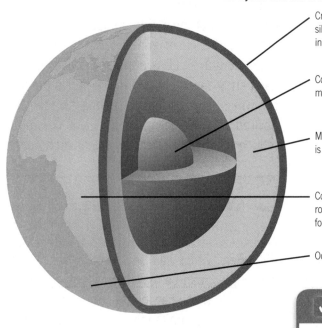

Crust: The crust is the outermost layer of the Earth and is rich in silicon, oxygen and aluminium. Much of the silicon and oxygen in the Earth's crust is present as silicon dioxide or silica.

Core: The core lies at the centre of the Earth. It is thought to be made of iron and nickel.

Mantle: The mantle is found between the crust and the core and is partially liquid. Rocks in the mantle flow slowly.

Continental crust: The continental crust consists of sedimentary rock, igneous rock and metamorphic rock. The igneous rock found in the continental crust is mainly granite.

Oceanic crust: The oceanic crust consists mainly of basalt.

✓ Maximise Your Marks

Make sure you can recall the names and details of the different layers of the Earth.

Evidence of the Structure of the Earth

The Earth's crust is too thick to drill through so evidence for the layered structure of the Earth comes from studies of the way that **seismic waves** (the shock waves sent out by earthquakes) travel through the Earth.

Build Your Understanding

The material through which the shock waves travel affects the speed of these waves. Studies show that the outer core of the Earth is liquid while the inner core is solid. The overall density of the Earth is greater than the density of the rocks that make up the Earth's crust. This means that the rocks in the mantle and the core must be much denser than the rocks we observe in the crust. Scientists believe that the core is mainly made of iron and nickel.

Movement of the Crust

Scientists used to believe that the features of the Earth's surface, such as the mountain ranges, were formed when the surface of the Earth shrank as it cooled down. However, scientists now believe that the Earth's geological features can be explained using a single, unifying theory called **plate tectonics**.

In 1914, the scientist Alfred Wegener first proposed **continental drift**, the idea behind plate tectonics. Initially, these ideas were resisted, particularly by religious groups. Scientists examining Wegener's theory could not, at first, explain how or why the plates moved, but as more evidence emerged, the theory of plate tectonics was gradually accepted.

Build Your Understanding

The main idea behind plate tectonics is that the Earth's lithosphere (the crust and upper mantle) is split up into 12 large plates. Each plate moves slowly over the Earth's surface at a rate of a few centimetres each year and is caused by convection currents in the mantle. These currents are caused by the natural radioactive decay of elements deep inside the Earth that release heat energy.

By studying geological processes, scientists have been able to explain what happened in the past. Scientists now believe that at one time all the continents were joined together to form a supercontinent called Pangea. Since that time, the continents have moved apart.

Evidence that Supports the Theory of Plate Tectonics

There are many clues that support the ideas about plate tectonics:
- When the South American coast was first mapped, people noticed that the east coast of South America and the west coast of Africa fitted together like pieces of an enormous jigsaw.
- The examination of fossil remains in South America and Africa showed that rocks of the same age contained the remains of an unusual freshwater crocodile-type creature.
- Further evidence that South America and Africa were once joined was uncovered when scientists discovered that rock strata of the same age were strikingly similar on both sides of the Atlantic.
- British rocks that were created in the **Carboniferous period** (300 million years ago) must have formed in tropical swamps. Yet rocks formed in Britain 200 million years ago must have formed in deserts. This shows that Britain must have moved through different climatic zones as the tectonic plate that Britain rests on moved across the Earth's surface.

? Test Yourself

1. What name is given to the outer layer of the Earth?
2. In which state is the Earth's inner core?
3. Which three elements are the most abundant in the Earth's crust?
4. Which elements are abundant in the Earth's core?

★ Stretch Yourself

1. What is the Earth's lithosphere?
2. a) Describe why the Earth's plates move.
   b) Where does our evidence for the structure of the Earth come from?

Consequences of Plate Tectonics

Plate Movements

The movement of tectonic plates causes many problems, including **earthquakes** and **volcanoes**. These tend to be worse near the edges of plates, known as the plate boundaries. The plates can move in three different ways:

- They can slide past or over each other.
- They can move towards each other.
- They can move away from each other.

These diagrams show how the Earth's plates can move

Earthquakes

Earthquakes are caused by tectonic plates sliding past or over each other. The San Andreas Fault in California is a famous example of where this occurs.

The plates in this area have fractured into a very complicated pattern. As the plates start moving past each other they tend to stick together, rather than slide smoothly past. When the plates stick together, forces build up until eventually the plates move suddenly. The strain that has built up is released in the form of an earthquake. If this happens beneath the oceans, it can result in catastrophic **tsunami** waves.

Scientists have studied earthquakes in an effort to predict when they will occur and so warn people to move away from the affected areas. However, with so many factors involved it is rarely possible to predict exactly when an earthquake or a volcanic eruption will occur. When they do happen, they can cause massive destruction and loss of life.

The San Andreas Fault

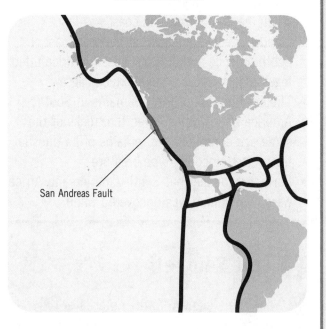

San Andreas Fault

Volcanoes

Like earthquakes, volcanoes are found in locations around the Earth where two plates are moving towards each other. By studying where most earthquakes and volcanoes happen, scientists have been able to identify plate boundaries.

These **convergent plate boundaries** often involve the collision between an oceanic and a continental plate (subduction zone). Oceanic plates contain minerals that are rich in the elements iron and magnesium, and are denser than continental plates. When an oceanic plate and a continental plate converge, the denser oceanic plate is forced beneath the continental plate. The continental plate is stressed, and the existing rocks are folded and metamorphosed.

As the oceanic plate is forced down beneath the continental plate, seawater lowers the melting point of the rock and some of the oceanic plate may melt to form **magma**. If the magma has a lower density than the surrounding rock in the Earth's crust it can rise up through weaknesses or cracks in the crust to form volcanoes.

Magma is molten rock below the Earth's surface; it becomes lava as it reaches the Earth's surface. Lava erupts from volcanoes. Some iron-rich basaltic lavas are runny and are relatively safe. Elsewhere, silica-rich viscous lavas are produced. These lavas are much more dangerous: they explode violently, often producing pumice, clouds of choking ash and throwing out pieces of rock called bombs.

When the lava cools down and solidifies, it forms igneous rocks. The faster the lava cools down the smaller the crystals in the rock will be.

Some people choose to live near to volcanoes, even though they might erupt, because the soils formed when the igneous rocks are weathered are very fertile.

A convergent plate boundary along the Western coast of South America is responsible for the formation of the Andes mountain range.

Build Your Understanding

Another consequence of plate tectonics is the formation of mid-ocean ridge basalts. When tectonic plates move apart, magma comes to the surface. This usually occurs under oceans. As the molten rock cools, it solidifies and forms the igneous rock basalt. These plate boundaries are often referred to as constructive plate boundaries because new crust is being made. Basalt is rich in iron, which is magnetic. As the basalt cools down, the iron-rich minerals in the basalt line up with the Earth's magnetic field. By examining the direction in which these minerals have lined up, scientists in the 1960s discovered that they could work out the direction of the Earth's magnetic field when the rocks crystallised.

However, examination of the basalt rocks on either side of a mid-ocean ridge shows a striped magnetic reversal pattern. The pattern is symmetrical about the ridge and provides evidence that the Earth's magnetic field periodically changes direction. This reversal appears to be very sudden and occurs about every half a million years. According to the rock record, another reversal is now well overdue!

? Test Yourself

1. What causes earthquakes?
2. Why can't scientists predict the exact date of an earthquake?
3. Why do earthquakes occur near volcanoes?
4. Why do oceanic plates move below continental plates when they collide?

★ Stretch Yourself

1. Explain how magnetic reversal patterns are formed near ocean ridges.

Everyday Chemistry

Earth and Pollution

Cooking

During a **chemical reaction**, a new substance is made in an **irreversible** process and there is an **energy change**. Cooking food is an example of a chemical reaction.

Eggs and meat are examples of **proteins**. When cooked, their appearance and texture change. Potatoes are a source of **carbohydrate**. When they are cooked, they become softer and fluffier and their taste improves.

When eggs or meat are cooked, the protein molecules change shape. This irreversible process is called **denaturing**.

When potatoes are cooked the starch grains become larger, the cell walls break open and the potato becomes softer and easier to digest.

Paint

Paint is used to make surfaces look more attractive and to protect them from damage. The paint is put on to the surface in thin layers. The solvent in the paint evaporates and the paint dries. Paint is a special type of mixture called a **colloid**. Paints consist of a solvent to thin the paint out, a pigment to give the colour and a binding medium to make the pigment stick to the surface it is coating. In a colloid, the particles of the pigment mix throughout the paint, but they are not dissolved in the solvent.

Phosphorescent pigments take in energy and then release it later as light. They are used in a range of applications such as watch dials.

Thermochromic pigments change colour as the temperature changes. They can be used to help monitor temperatures, for example on bath toys, cups and cutlery for babies. Parents can see at a glance if the temperature is safe for their baby.

Build Your Understanding

Colloids do not separate into layers because the particles are so small that they are fully dispersed throughout the mixture and do not settle completely to the bottom over time. Oil paints consist of pigment particles dispersed in oil and a solvent. The solvent evaporates and the oil is oxidised by oxygen in the air.

Washing-up Liquid

Each ingredient in washing-up liquid has a specific function to help the cleaning process:
- **Detergent**, to clean the object being washed.
- Water, to thin out the detergent so it is easy to dispense.
- Colouring and fragrance, to make it more appealing to users.
- A rinse agent, to help the water run off the washed objects more easily.

? Test Yourself

1. What happens during a chemical reaction?
2. Why are colourings and fragrances added to washing-up liquids?
3. What is special about thermochromic pigments?

★ Stretch Yourself

1. Explain why colloids do not separate over time.

The Carbon Cycle

Moving Carbon Around the Carbon Cycle

The level of carbon dioxide in the atmosphere is fairly constant. This is because of the **carbon cycle**. This cycle moves carbon between the atmosphere, the oceans and rocks.

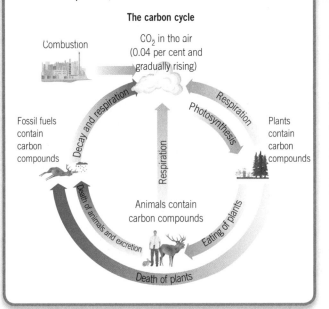

The carbon cycle

Combustion

CO_2 in the air (0.04 per cent and gradually rising)

Respiration

Photosynthesis

Respiration

Decay and respiration

Fossil fuels contain carbon compounds

Plants contain carbon compounds

Death of animals and excretion

Animals contain carbon compounds

Eating of plants

Death of plants

Build Your Understanding

- Plants take in carbon from the atmosphere during photosynthesis.
- Plants and animals return carbon to the atmosphere during respiration.
- Fossil fuels release carbon into the atmosphere during combustion (burning).

Properties of a Good Fuel

A good fuel should be:
- Easy to store and transport.
- Inexpensive to buy.
- Non-polluting.
- Easy to burn, producing little ash or smoke.
- Non-toxic.
- Widely available.
- Efficient when it is burned (produce lots of energy).

⚠ Boost Your Memory

Make a poster to show how carbon is moved between the atmosphere, oceans and rocks. Make sure you annotate each stage.

Fossil Fuels

As the world's population expands, and with greater worldwide industrialisation, the overall amount of fossil fuels being burned is increasing. This means more carbon dioxide is being released into the atmosphere.

Fossil fuels such as coal, oil and gas are very convenient to use.

❓ Test Yourself

1. Where is carbon found?
2. Why is more carbon dioxide now being released into the atmosphere?
3. Name three fossil fuels.
4. How do plants take in carbon?
5. What is happening as a result of the expansion in the world's population?

⭐ Stretch Yourself

1. Carbon moves between the atmosphere, the oceans and rocks through a series of steps called the carbon cycle. Explain why the levels of carbon dioxide in the atmosphere have remained almost constant until recent times.

Practice Questions

Complete these exam-style questions to test your understanding. Check your answers on page 121. You may wish to answer these questions on a separate piece of paper.

1 This question is about how the amount of oxygen in the atmosphere has increased over time. Use the words and phrases below to complete the sentences. (4)

- **Carbon dioxide**
- **Harmful**
- **Ozone**
- **Plants**

As ...... **a** ...... evolved, the amount of oxygen in the atmosphere increased. These plants

grew well in the ...... **b** ...... -rich atmosphere. As the amount of oxygen increased, an ...... **c** ...... layer

developed. This layer filtered out ...... **d** ...... ultraviolet rays.

2 Over time, the composition of the Earth's atmosphere has changed. Scientists believe that the Earth is 4.5 billion years old. During the first billion years of the Earth's history there was great volcanic activity. Volcanoes released large amounts of gas, which formed the Earth's early atmosphere.

a) Which gas was not produced in large quantities by the volcanoes? Tick one box.

☐ Methane   ☐ Carbon dioxide

☐ CFCs   ☐ Steam

☐ Ammonia   (1)

b) The Earth's early atmosphere mainly consisted of carbon dioxide. Name a planet in the solar system that has an atmosphere similar to the Earth's early atmosphere. (1)

....

c) What did the steam released by the volcanoes eventually produce? (1)

....

d) Why did the amount of oxygen in the atmosphere eventually increase? (1)

....

e) The amount of nitrogen in the Earth's atmosphere has also increased over time. Some nitrogen was produced when ammonia reacted with oxygen. How else was nitrogen produced? (1)

....

3 Many fuels contain carbon. Complete the equation to show what happens when carbon burns to form carbon dioxide.

a) $C + \underline{\hspace{1cm}} \rightarrow CO_2$ (1)

b) Humans are affecting the proportion of gases in the atmosphere.

i) How is the amount of carbon dioxide in the atmosphere changing? (1)

....

ii) Why is the amount of carbon dioxide in the atmosphere changing? (1)

....

c) Some fuels contain traces of sulfur. Complete the equation to show what happens when sulfur burns to form sulfur dioxide.

$S + O_2 \rightarrow$ _____ (1)

d) Which of these environmental problems could be caused by acid rain? Tick two boxes. (2)

- [ ] Damage to statues
- [ ] Skin cancer
- [ ] Global dimming
- [ ] Changes to weather patterns
- [ ] Damage to trees
- [ ] Increased sea levels

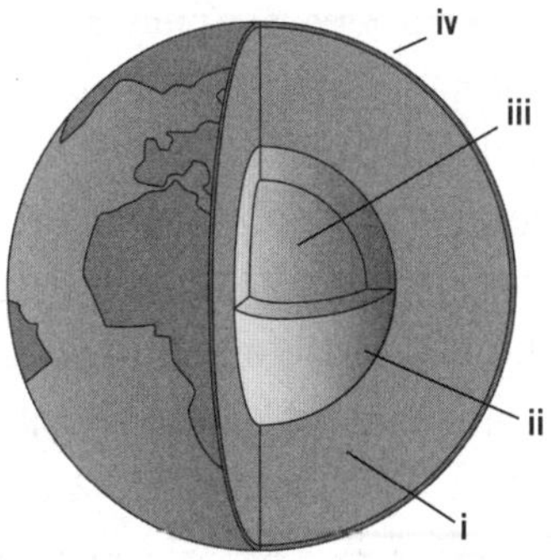

4 This diagram shows the layered structure of the Earth.

a) Add the missing labels to complete the diagram. (4)

b) The crust and upper mantle are split into a number of moving plates. What type of rock is formed when plates are stressed at a plate boundary? (1)

....

c) Sometimes an oceanic plate and a continental plate collide. Why is the oceanic plate forced beneath the continental plate? (1)

....

d) When an oceanic plate and a continental plate collide, mountain ranges can be formed. Where is this process forming mountain ranges today? Tick one box. (1)

- [ ] Iceland
- [ ] California
- [ ] West coast of South America
- [ ] West coast of Africa

e) Give an example of a natural disaster that is associated with problems at plate boundaries. (1)

....

f) Why can scientists not predict exactly when an earthquake will occur? (1)

....

5 Aluminium is extracted from its ore, bauxite. Aluminium is used to make drinks cans. Once used, the drinks cans can be recycled or buried in landfill sites. Use your knowledge and understanding to explain why people should be encouraged to recycle aluminium cans.

You should make sure your answers are written using good spelling, punctuation and grammar. (6)

....

....

....

....

....

....

How well did you do?

0–9 Try again	10–18 Getting there	19–25 Good work	26–30 Excellent!

Organic Chemistry 1

The Importance of Carbon

Carbon atoms have the ability to form four bonds with other atoms. This means that carbon atoms can be made into a large number of different compounds. These compounds are the basis of life and the chemistry of these compounds is called **organic chemistry**. Organic compounds contain covalent bonds.

✓ Maximise Your Marks

Covalent bonding involves the sharing of pairs of electrons. The shared pairs of electrons hold the atoms together.

Alkanes

Alkanes are a family of hydrocarbon molecules. Alkanes are hydrocarbons, so only contain hydrogen and carbon atoms. Most of the compounds in crude oil are hydrocarbons.

Scientists describe alkanes as **saturated** hydrocarbons. This is because they contain no carbon–carbon double bonds (C=C) and so contain the maximum number of hydrogen atoms.

The alkanes are an example of an **homologous series**: all alkanes have the same general formula and similar chemical properties.

Their physical properties, for example boiling points, vary gradually down the series.

Name	Methane	Ethane	Propane	Butane
Chemical Formula	CH_4	C_2H_6	C_3H_8	C_4H_{10}
Structure	H | H–C–H | H	H H | | H–C–C–H | | H H	H H H | | | H–C–C–C–H | | | H H H	H H H H | | | | H–C–C–C–C–H | | | | H H H H

Build Your Understanding

Alkanes have the general formula C_nH_{2n+2}. Ball and stick models are a useful way of showing the three-dimensional position of atoms and bonds.

The ball and stick model of methane opposite shows each carbon atom forms bonds with four hydrogen atoms.

Ball and stick model of methane

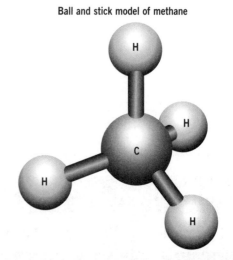

Build Your Understanding (cont.)

The carbon chain for some alkanes can be either branched or unbranched and this can cause isomerism.

The isomers of C_4H_{10} are shown below.

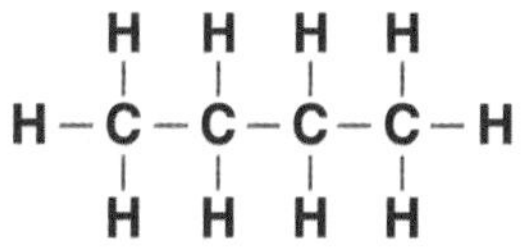

$CH_3(CH_2)CH_3$

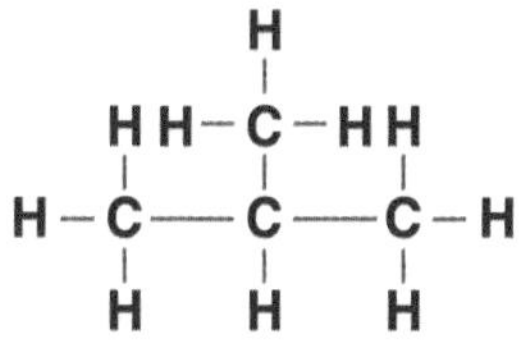

$(CH_3)_2CHCH_3$

The isomers of C_5H_{12} are shown below.

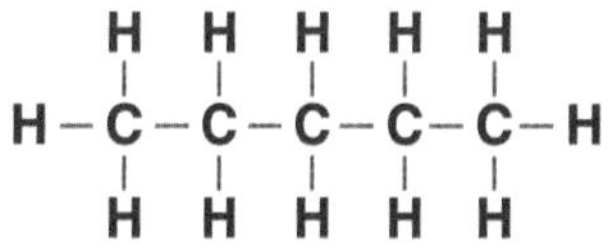

$CH_3(CH_2)_3CH_3$

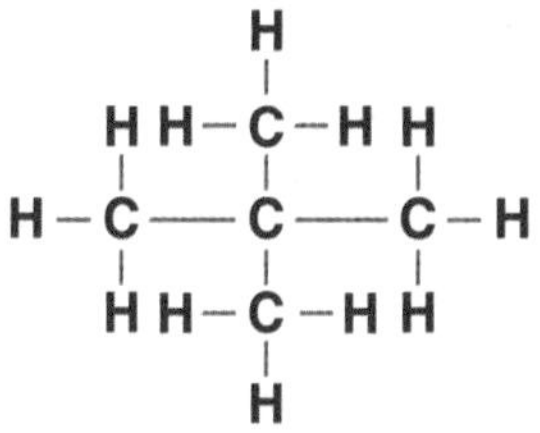

$C(CH_3)_4$

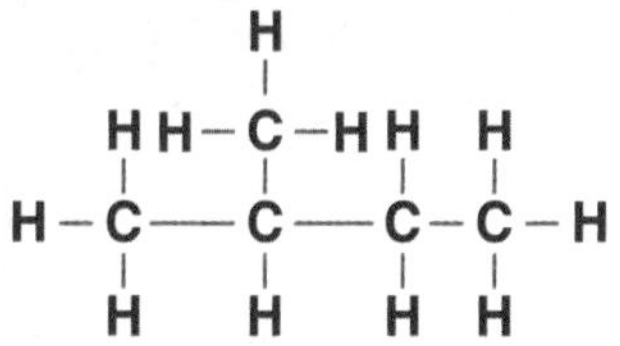

$(CH_3)_2CHCH_2CH_3$

Isomers

Isomers are molecules that have the same number and kind of atoms, but these atoms are arranged in different ways. Isomers have the same molecular formula, but a different structural formula and display formula.

Alkenes

Alkenes are also hydrocarbon molecules. Scientists describe alkenes as **unsaturated** hydrocarbons because they all contain one or more C=C bond. They are produced by cracking longer chain alkane molecules.

Alkenes are more reactive than alkanes owing to the presence of C=C bonds. This means alkenes are more useful because they can be used to make new substances.

Alkenes have the general formula C_nH_{2n}.

Alkenes react with bromine water. Bromine water is decolourised in the presence of alkenes. When alkenes react with bromine water an addition reaction occurs (the two molecules add together to form one molecule).

Bromine atoms are added to each of the carbon atoms involved in the double bond. This forms a colourless **dibromo** compound.

? Test Yourself

❶ How many bonds do carbon atoms form?

❷ Why are alkanes described as saturated hydrocarbons?

❸ What is the name of the first member of the alkane family?

❹ What is the molecular formula of butane?

★ Stretch Yourself

❶ Consider these compounds:
- $C_{56}H_{112}$
- C_2H_5OH
- $C_{18}H_{38}$

a) Which of these compounds belongs to the alkene family?

b) Which of these compounds belongs to the alkane family?

Fuels

Formation of Fossil Fuels

Fuels are burned to release energy. In the UK, the fossil fuels coal, oil and natural gas are widely used. The burning of fuels is an **exothermic** reaction. The products of crude oil fractional distillation can be used to make a wide range of useful materials. Fossil fuels are formed over millions of years from the fossilised remains of dead plants and animals. When the plants and animals died, they fell to the sea or swamp floor.

Occasionally, the remains were covered very quickly by **sediment**. In the absence of oxygen, the remains did not decay. Over time, more layers of sediment gradually built up. The lower layers became heated and **pressurised** forming fossil fuels. Fossil fuels are **non-renewable**, so crude oil is a **finite** resource. They take millions of years to form, but are being used up very quickly.

Crude Oil

Crude oil is a mixture of many substances, but the most important are hydrocarbons (molecules that only contain hydrogen and carbon atoms).

Some of the hydrocarbons have very short chains of carbon atoms. These hydrocarbons:
- Are less **viscous** (more runny).
- Are easy to **ignite**.
- Have low boiling points.
- Are valuable fuels.

Longer hydrocarbon molecules are less useful as fuels. However, before any of these hydrocarbons can be used, they must first be separated into groups of molecules with a similar number of carbon atoms called **fractions**.

In compounds, the atoms of two or more different elements are chemically combined. In mixtures, two or more different elements, or compounds, are simply mixed together. Each constituent part of the mixture has its original chemical properties. This makes it quite easy to separate mixtures.

Crude oil is a mixture so it is possible to separate it into its different parts.

> ✓ **Maximise Your Marks**
>
> Coal is mainly carbon. Petrol, diesel and oil are hydrocarbons.

Fractional Distillation of Crude Oil

The components of crude oil can be separated by fractional distillation. First, the crude oil is heated until it eventually **vaporises**. The fractionating column is much hotter at the bottom than at the top. This means that short hydrocarbon molecules can reach the top of the column before they **condense** and are collected. Longer hydrocarbon molecules condense at higher temperatures and are collected lower down the column.

No. carbon atoms in hydrogen chain	Temperature	Fraction collected
3	less than 40 °C	refinery gas
8	40–205 °C	petrol
10	60–110 °C	naphtha
15	175–325 °C	kerosene
20	250–350 °C	diesel
35	300–600 °C	oil
50+	600 °C	bitumen

Forces of Attraction

The forces of attraction *between* the hydrocarbon molecules are much weaker than the forces of attraction *within* the molecules. The larger the hydrocarbon molecule, the stronger the forces of attraction between the molecules, so more energy is required to overcome them. Larger-molecule hydrocarbons have higher boiling points.

The different fractions obtained from crude oil have different uses:
- Liquid petroleum gas (LPG) contains propane and butane. These gases are used in domestic heating and cooking.
- Petrol is used as a fuel for cars.
- Kerosene is used as a fuel for aeroplanes.

- Diesel oil is used as a fuel for cars, lorries and trains.
- Fuel oil is used as a fuel for large ships and power stations.
- Bitumen is used to make roads and roofs.

✓ Maximise Your Marks

Remember, in the fractional distillation of crude oil, the crude oil mixture is first vaporised. The mixture is then separated because the different fractions condense at different temperatures. Do not confuse fractional distillation with cracking or with the blast furnace.

Cracking

The large-molecule hydrocarbons separated during the fractional distillation of crude oil are not very useful. However, they can be broken down into smaller, more useful and more valuable molecules by a process called **cracking**.

Industrial Cracking
The cracking of long chain hydrocarbons is carried out on a large scale. First, the long-molecule hydrocarbon is heated until it vaporises. The vapour is then passed over a hot aluminium oxide catalyst.

Build Your Understanding

Octane is one of the hydrocarbon molecules in petrol. Ethene, a member of the alkene family of hydrocarbons, is also produced. Ethene is used to make a range of new compounds including plastics and industrial alcohol. As we only have a finite amount of crude oil left, scientists are working to find replacement fuels for the future.

Make sure you can sketch the apparatus for the cracking of liquid paraffin in the laboratory.

✓ Maximise Your Marks

Make sure that when you balance equations for cracking you have the same number of each type of atom on both sides of the equation.

? Test Yourself

1. How long does it take for fossil fuels to form?
2. Which elements are found in hydrocarbon molecules?
3. Give three properties of short chain hydrocarbon molecules.

★ Stretch Yourself

1. Describe why larger-molecule hydrocarbons have higher boiling points than smaller hydrocarbons.

Organic Chemistry and Analysis

Vegetable Oils

Extracting Vegetable Oils

Plant oils are a valuable source of energy in our diets and are also essential sources of vitamins A and D. If we eat too many vegetable oils, however, we could suffer from health problems, such as heart disease, in later life.

Vegetable oils can be **extracted** from the fruits, seeds or nuts of some plants. Popular vegetable oils include olive oil, rapeseed oil, lavender oil and sunflower oil. Living things make fats and oils to store energy.

Vegetable oils are often removed by crushing up the plant material and collecting the oil. Other oils are collected using **distillation**. These processes remove water and other impurities to produce pure oil.

Fats have higher boiling points than water. Cooking food by frying is therefore much faster than cooking food in boiling water. In addition, frying foods produces interesting new flavours and increases the energy content of the food.

Fuels and Soaps

When vegetable oils are burned they release lots of energy. In fact, vegetable oils can be used in place of fossil fuels in the form of **biodiesel**. This is an alternative to diesel produced from crude oil. Fuels made from plant materials are called **biofuels**. Oils and fats are **esters**.

Soap can be produced by reacting vegetable oils or animal fats with hot sodium or potassium hydroxide solution. The process is known as **saponification** and also produces glycerol. Soaps are the sodium or potassium salts of carboxylic acids with long carbon chains.

✓ Maximise Your Marks

Fat + Sodium Hydroxide → Soap + Glycerol

This is a **hydrolysis** reaction.

Emulsions

Oils do not dissolve in water. Salad dressing is an example of a type of everyday mixture called an **emulsion**. It is a mixture of two liquids: oil and water containing the vinegar. Salad dressing is made by shaking oil and vinegar so that they mix together. After a short while, however, the oil and vinegar separate out to form two distinct layers. If the salad dressing was placed in a separating funnel, the lower, denser layer could be run off leaving the less dense layer in the separating funnel. Many of the salad dressings bought from shops contain molecules called **emulsifiers** that help the oil and vinegar to stay mixed together. An emulsion is thicker than either of its separate parts. Emulsions have some special properties that make them very useful. Emulsions can improve a product's texture, appearance or ability to coat foodstuffs. Milk is an oil-in-water emulsion. Butter is a water-in-fat emulsion.

Build Your Understanding

Emulsifiers are molecules with two very different ends. In a salad dressing, one end of the emulsifier molecule is attracted to the oil, while the other end is attracted to the water in the vinegar. The addition of emulsifier molecules keeps the two liquids mixed together.

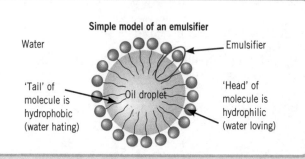

Simple model of an emulsifier

Water

Emulsifier

'Tail' of molecule is hydrophobic (water hating)

Oil droplet

'Head' of molecule is hydrophilic (water loving)

Saturated Fats

Animal fats are usually solid, or nearly solid, at room temperature. A **saturated fat** contains many C–C bonds, but no C=C bonds. Scientists believe that people who eat lots of saturated fats may develop raised blood cholesterol levels. This is linked with an increased risk of heart disease. Most vegetable fats are liquids at room temperature and so are described as oils.

Unsaturated Fats

Vegetable oils contain C=C bonds. Scientists describe these molecules as **unsaturated fats** because they could bond to more hydrogen atoms. The presence of the C=C bonds affects the way that the fatty acid in molecules can pack together. C=C bonds are rigid and their presence causes kinks in the carbon chain so that the fatty acids cannot pack closely together. Unsaturated fats have lower melting points than saturated fats. While most vegetable fats are liquid at room temperature, most animal fats are solids. Bromine water can be used to detect the presence of the C=C bonds in vegetable oils.

Hydrogenated Vegetable Oils

Vegetable oils are often liquids at room temperature because they contain C=C bonds. There are, however, advantages to using fats that are solid at room temperature. They are easier to spread and can be used to make new products such as cakes, pastries and spreads.

Vegetable oils can be made solid at room temperature by a process known as **hydrogenation**. The oils are heated to 60 °C with hydrogen and a nickel catalyst. Hydrogen adds to the C=C bonds.

> ✓ **Maximise Your Marks**
>
> Hydrogenation produces saturated fats, which is harmful to health.

Food Additives

Scientists often add chemicals to improve foods. The chemicals that have passed safety tests and are approved for use throughout the European Union are given **E-numbers**. Chemicals that are commonly added to foods include:

- **Colours**, which are added to make food look more attractive.
- **Flavours**, which are added to enhance taste.
- **Antioxidants**, which stop foods from reacting with oxygen.

> ✓ **Maximise Your Marks**
>
> Make sure you are able to say why each of these common additives are put in foodstuffs.

Organic Chemistry and Analysis

❓ Test Yourself

1. From which parts of plants can we obtain oils?
2. What are biofuels?
3. Which vitamins do we obtain from eating fats?
4. Why are antioxidants added to foods?

⭐ Stretch Yourself

1. A new fat is found that is liquid at room temperature.
   a) Is this fat likely to be saturated or unsaturated?
   b) Is this fat likely to have come from a plant or an animal?
   c) How could you test if this fat was saturated or unsaturated?

Plastics

Polymerisation

Plastics are **synthetic** (manufactured) **polymers**. **Natural polymers** include cotton, wood, leather, silk and wool. In polymers, lots of small molecules are joined together to make one big molecule.

The simplest alkene, ethene, can be formed by the cracking of large hydrocarbon molecules. If ethene is heated under pressure in the presence of a **catalyst**, many ethene molecules can join together to form a compound composed of large molecules called poly(ethene) or polythene. The diagram shows how a large number of ethene molecules join together to form polythene.

$$n \quad \begin{array}{c} H \quad H \\ | \quad\ | \\ C = C \\ | \quad\ | \\ H \quad H \end{array} \rightarrow \left(\begin{array}{c} H \quad H \\ | \quad\ | \\ C - C \\ | \quad\ | \\ H \quad H \end{array} \right)_n$$

The 'n' at the start of the equation and the section of the polymer surrounded by brackets represents the number of molecules involved. The brackets are used because it would be impractical to write out the complete structure. The brackets surround a representative unit that is then repeated through the whole polymer.

The small starting molecules, in this case the ethene molecules, are called **monomers**. The ethene molecules join together to form long chain molecules called polymers. A polymer is made from lots of monomer units.

The polythene diagram is an example of an **addition polymerisation** reaction. The ethene molecules have simply joined together.

Other Polymers

Polymerisation reactions may involve other monomer units. The exact properties of the polymer formed depend upon:
- The monomers involved.
- The conditions under which it was made.
- The length of the polymer chains.

Low density polythene and high density polythene have very different properties and uses because they are produced using different catalysts and different conditions. A plastic's properties are also affected by the **crystallinity** of its structure; the more crystalline a plastic is the more brittle it will be.

A crystalline polymer

Regular polymer chains closely packed.

Build Your Understanding

Polypropene is made by an addition polymerisation reaction between many propene molecules.

$$n \quad \begin{array}{c} CH_3 \quad H \\ | \quad\quad\ | \\ C = C \\ | \quad\quad\ | \\ H \quad\ H \end{array} \rightarrow \left(\begin{array}{c} CH_3 \quad H \\ | \quad\quad\ | \\ C - C \\ | \quad\quad\ | \\ H \quad\ H \end{array} \right)_n$$

Polyvinyl chloride (PVC) is made by an addition polymerisation reaction between many chloroethene molecules. Chloroethene used to be called vinyl chloride.

$$n \quad \begin{array}{c} Cl \quad H \\ | \quad\ | \\ C = C \\ | \quad\ | \\ H \quad H \end{array} \rightarrow \left(\begin{array}{c} Cl \quad H \\ | \quad\ | \\ C - C \\ | \quad\ | \\ H \quad H \end{array} \right)_n$$

Polytetrafluoroethene (PTFE or Teflon™) is made by an addition polymerisation reaction between many tetrafluoroethene molecules.

$$n \quad \begin{array}{c} F \quad F \\ | \quad\ | \\ C = C \\ | \quad\ | \\ F \quad F \end{array} \rightarrow \left(\begin{array}{c} F \quad F \\ | \quad\ | \\ C - C \\ | \quad\ | \\ F \quad F \end{array} \right)_n$$

PTFE is known as 'Teflon'. Surfaces coated in Teflon™ have low friction. It is used to coat some frying pans and saucepans.

Thermoplastics and Thermosetting Plastics

Thermoplastics consist of long polymer chains with few cross-links. When heated, these chains untangle and the material softens. It can then be reshaped. On cooling, the material becomes solid and stiff again. Thermoplastics can be heated and reshaped many times. Polythene is a thermoplastic. Thermoplastics can be stretched easily.

Thermosetting plastics consist of long, heavily cross-linked polymer chains. When they are first made these thermosetting plastics are soft and can be shaped. Once they have set, however, they become solid and stiff. They do not soften again, even if they are heated to very high temperatures, and so they cannot be reshaped. Thermosetting plastics, such as melamine, are rigid and cannot be stretched.

Nylon and Polyester

Nylon and polyester are condensation polymers that are used to make clothes. Nylon can be used to make cheap, waterproof jackets. Nylon is lightweight, hard wearing, keeps UV light out and is waterproof, but it is not breathable so perspiration can make a nylon jacket quite uncomfortable to wear.

Build Your Understanding

More expensive Gore-tex™ jackets are breathable. Gore-tex™ consists of a thin membrane of PTFE, which is used to coat nylon fabrics. The membrane has lots of little holes. Liquid water is too big to go through these holes, so the fabric is waterproof. Water vapour is small enough to pass through the holes, so it is breathable.

Uses of Polymers

Polymers have some very useful properties. They are flexible, good thermal and electrical insulators, resistant to corrosion, waterproof and are easy to mould and shape as they generally have low melting points:

- **Polythene** is cheap and strong. It is used to make plastic bags and bottles. Polythene bags are cheaper than the paper bags they have widely replaced.
- **PVC** is rigid and can be used for building materials such as drainpipes. PVC has replaced metal drainpipes because PVC is cheaper and lighter than metal. Chemicals called **plasticisers** can be added to PVC to make products such as Wellington boots and mackintoshes.
- **Polypropene** is strong and has a high elasticity. It is used for crates and ropes.
- **Polystyrene** is cheap and can be moulded into different shapes. It is used for packaging and for plastic casings.

? Test Yourself

1. What type of reaction is involved in the formation of polythene?
2. What is the name given to the small units that join together to form a polymer?
3. Suggest two uses of the polymer polythene.

★ Stretch Yourself

1. Draw a tetrafluoroethene molecule.
2. Is tetrafluoroethene saturated or unsaturated? Explain your answer.
3. Draw the repeating unit for the polymer made from the addition polymerisation of tetrafluoroethene molecules.

Ethanol

Ethanol

Ethanol, C_2H_5OH, is a member of the **alcohol** family of organic compounds. The diagram shows the structure of ethanol. It is not a hydrocarbon because it contains an oxygen atom as well as carbon and hydrogen atoms.

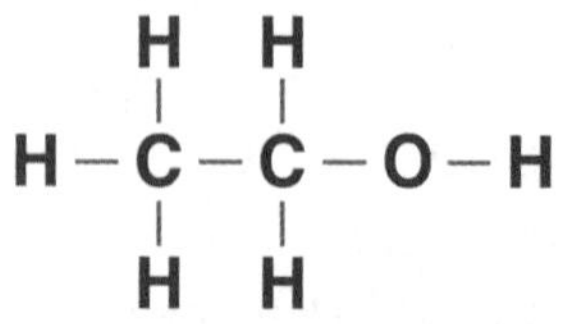

$$
\begin{array}{ccc}
\text{H} & \text{H} & \\
| & | & \\
\text{H}-\text{C}-\text{C}-\text{O}-\text{H} \\
| & | & \\
\text{H} & \text{H} &
\end{array}
$$

Ethanol as a Fuel

In some countries, sugar from sugar beet or sugar cane is made into alcohol. This alcohol is then mixed with petrol to produce a fuel for vehicles. Ethanol is a **renewable** energy resource that burns very cleanly, producing carbon dioxide and water vapour. Alcohols, however, release less energy than petrol when they are burned. In order to produce enough alcohol for fuel, large areas of fertile land and a favourable climate are required to grow the plants needed. Producing fuels in this way is particularly attractive for countries that do not have large reserves of fossil fuels. However, if land is being used to grow crops to produce ethanol, this will reduce the amount of land available to grow food for people to eat.

Build Your Understanding

Ethanol is found in drinks like beer and wine. It is, however, toxic in large amounts. Ethanol has many useful properties. It is a good solvent and evaporates quickly. Many aftershaves contain ethanol. Ethanol is an important raw material and can also be used as a fuel and as a biofuel.

Methanol is another member of the alcohol group and is even more toxic than ethanol. If someone was to drink methanol, they could become blind or even die. Methylated spirit is a mixture of ethanol, methanol and a purple dye. The purple dye is used to warn people about its toxicity; its unpleasant taste is to prevent people from drinking it. Methanol is a good solvent.

The Fire Triangle

A fire triangle shows the three things that must be present to produce a fire: oxygen, heat and fuel. If any one of these things is taken away, the fire will go out.

Fermentation

Fermentation has been used to make alcohol and alcoholic drinks for thousands of years. Fruits, vegetables and cereals are all sources of the sugar glucose, $C_6H_{12}O_6$.

During fermentation, yeast is used to **catalyse** (speed up) the reaction in which glucose is converted into ethanol and carbon dioxide:

$$\text{Glucose} \xrightarrow{\text{yeast}} \text{Ethanol} + \text{Carbon Dioxide}$$

✓ Maximise Your Marks

Make sure you can give the symbol equation for fermentation:

$$C_6H_{12}O_6 \rightarrow 2C_2H_5OH + 2CO_2$$

Fermentation (cont.)

The temperature of the fermentation reaction has to be carefully controlled. If the temperature falls too low, the yeast becomes less active and the rate of the reaction slows down. If the temperature rises too high, the yeast is denatured and stops working altogether. Temperatures between 25 °C and 50 °C work best. Water also needs to be present. Yeast is an **enzyme** or biological catalyst. It speeds up the **conversion** of sugar to alcohol and carbon dioxide, but is not itself used up in the process. Ethanol can also be made from waste biomass (biological material from living organisms) by genetically modified E. coli bacteria. The optimum conditions are:

- A pH of 6.0 to 7.0.
- A good supply of oxygen.
- A warm temperature of around 30 °C.

Vinegar is a solution that contains **ethanoic acid**. It is a weak acid so, compared with the same concentration of a strong acid like sulfuric acid, ethanoic acid will have a higher pH and will react more slowly with metals and metal carbonates.

Ethanol produced by fermentation has a concentration of around 6–14 per cent. Some people prefer drinks with a higher alcohol content, such as whisky and brandy. These higher concentrations of alcohol are achieved by the process of fractional distillation. There are health risks associated with drinking alcohol, and there is also an increased risk of accidents and raised levels of crime. Some religions prohibit the consumption of all alcoholic drinks.

Fermentation

The fermentation lock allows carbon dioxide to escape, but stops oxygen in the air from reacting with the alcohol. This is important as ethanol can easily be oxidised by microbes to form ethanoic acid, which would make the drink taste sour.

Industrial Alcohol

There is another, more modern, way of producing vast amounts of very pure alcohol. Ethene (which is produced during the cracking of long chain hydrocarbons) is reacted with steam to produce ethanol:

Ethene + Steam → Ethanol

Build Your Understanding

The symbol equation to sum up the production of industrial alcohol is:

$$C_2H_4 + H_2O \rightarrow C_2H_5OH$$

A catalyst of phosphoric acid and a temperature of 300 °C are used. This method of producing ethanol is much cheaper than fermentation. It is a continuous, rather than a batch, process and because the reaction between ethene (derived from crude oil, a fossil fuel) and steam is an addition reaction it has a higher atom economy than fermentation. Our reserves of fossil fuels are, however, finite and will run out one day.

❓ Test Yourself

1 What is the formula of ethanol?

2 Why is ethanol not a hydrocarbon?

3 Which crops can be used to produce sugar for making alcohol?

4 What is the catalyst used in fermentation?

⭐ Stretch Yourself

1 During fermentation reactions the temperature must be carefully controlled. What happens if the temperature falls too low or rises too high?

Organic Chemistry 2

Alcohols

The term 'alcohol' is often used for the compound ethanol. Alcohols contain the **hydroxyl**, OH, functional group and have the general formula $C_nH_{2n+1}OH$.

Each member of the family differs from the previous one by the addition of a CH_2 group. Alcohols are **neutral** and react with **carboxylic acids** to form **esters**.

Methanol CH_3OH	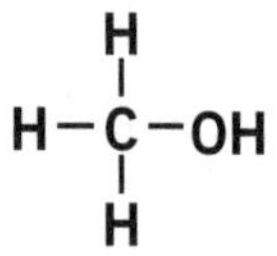 H | H—C—OH | H
Ethanol C_2H_5OH	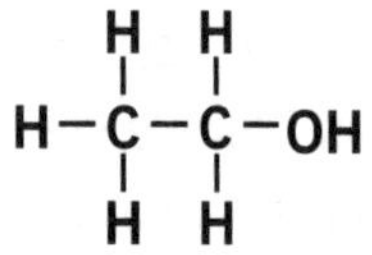 H H | | H—C—C—OH | | H H
Propanol C_3H_7OH	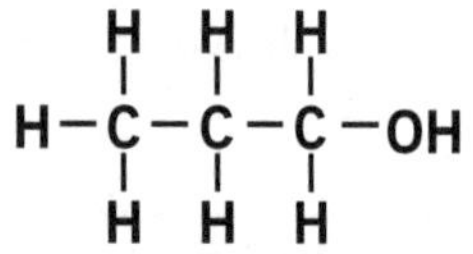 H H H | | | H—C—C—C—OH | | | H H H

Positional Isomers

Positional isomers are formed when a functional group can be placed in different positions along the carbon chain. The positional isomers of C_3H_8O are shown below:

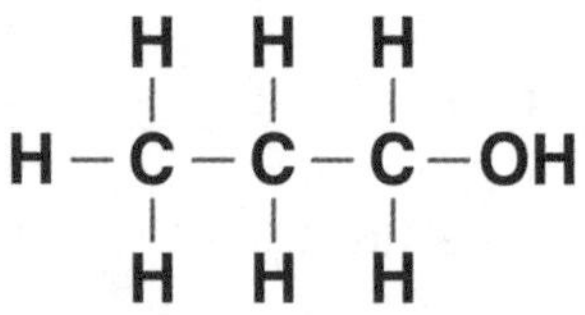

H H H
\| \| \|
H—C—C—C—OH $CH_3CH_2CH_2OH$ 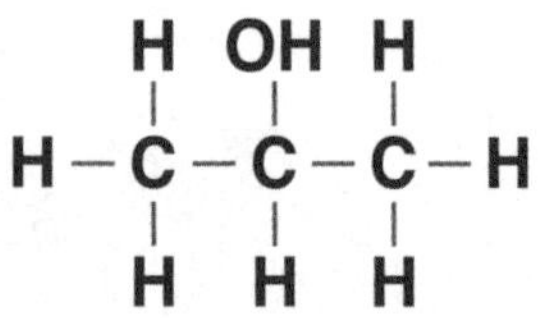
\| \| \|
H H H

H OH H
\| \| \|
H—C—C—C—H $CH_3CHOHCH_3$
\| \| \|
H H H

Uses of Alcohols

Alcohols have a wide range of uses:
- They are good solvents for many compounds.
- They can be burned and used as fuels. The alcohol can either replace petrol or be mixed with petrol to improve **combustion**.
- Ethanol is the alcohol found in alcoholic drinks such as beer.

Build Your Understanding

Methanol, ethanol and propanol all dissolve in water to form neutral solutions. The larger the alcohol molecule the less soluble in water it becomes. Alcohols react with sodium metal to form hydrogen gas and alkoxides:

Sodium + Ethanol → Hydrogen + Sodium Ethoxide

The larger the alcohol molecule the more slowly it reacts with sodium.

Water also reacts with sodium to form sodium hydroxide and hydrogen.

Alkanes are saturated hydrocarbons and so do not react with sodium. Alcohols and alkanes are both families of organic compounds.

The boiling points of alcohols are much higher than the boiling points of alkanes, which have the same number of carbon atoms, because there are stronger forces of attraction between the alcohol molecules than there are between the alkane molecules.

Alcohols can be burned in air. Complete combustion of ethanol produces carbon dioxide and water vapour. The alcohol is completely oxidised in this reaction:

Ethanol + Oxygen → Carbon Dioxide + Water Vapour

$$C_2H_5OH + 3O_2 \rightarrow 2CO_2 + 3H_2O$$

Build Your Understanding (cont.)

Ethanol can also be oxidised more gently by microbes or by chemical oxidising agents such as acidified potassium dichromate (VI). Primary alcohols can be oxidised to carboxylic acids. Ethanol is oxidised to ethanoic acid. Over time, the ethanol in an opened bottle of an alcoholic drink will react with the oxygen in the air to form ethanoic acid, leaving the drink with a distinctive vinegary smell and taste. Vinegar is an aqueous solution that contains ethanoic acid.

Carboxylic Acids

Methanoic acid HCOOH	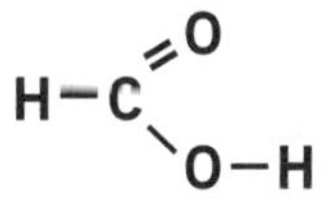
Ethanoic acid CH_3COOH	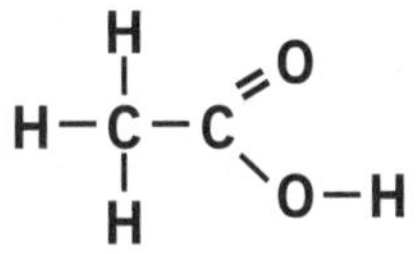
Propanoic acid C_2H_5COOH	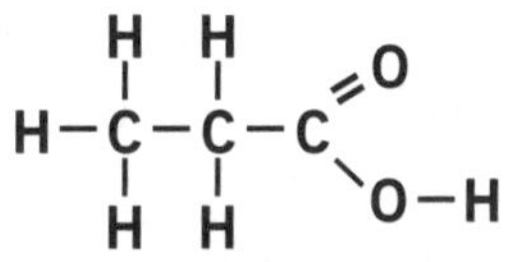

Carboxylic acids are described as weak acids. They can be identified by adding a few drops of universal indicator to a sample. Carboxylic acids will turn the indicator orange–red. Carboxylic acids react with metal carbonates to produce a salt, water and carbon dioxide:

Ethanoic Acid + Sodium Carbonate → Sodium Ethanoate (salt) + Water + Carbon Dioxide

Carboxylic acids react with alcohols to produce esters and water. The reaction is catalysed by strong acids:

Ethanoic + Ethanol ⇌ Ethyl Ethanoate + Water
Acid

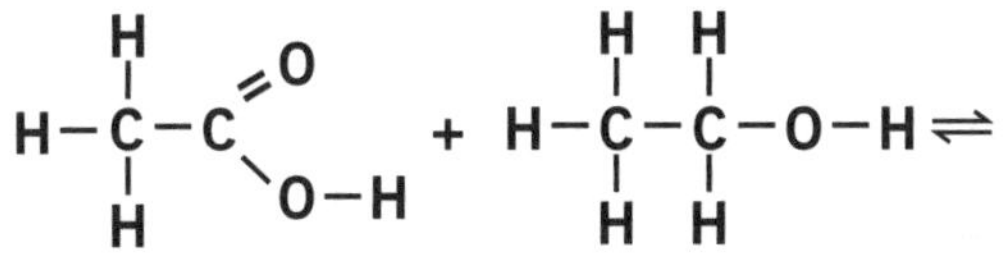

CH_3COOH CH_3CH_2OH

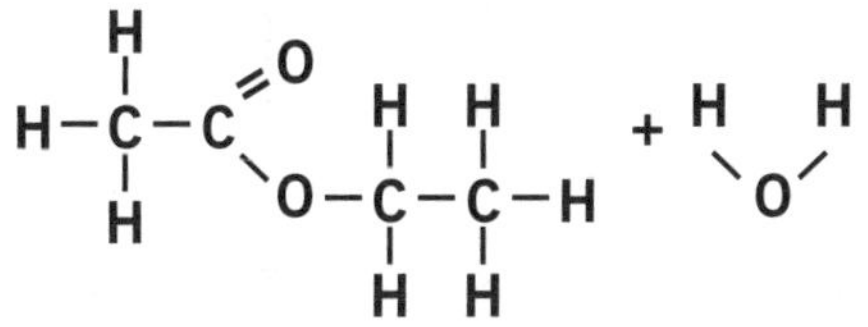

$CH_3COOCH_2CH_3$ H_2O

Strong and Weak Acids

Carboxylic acids are acids because they are proton (hydrogen ion) donors.

Carboxylic acids are weak because when they are dissolved in water they do not ionise (split up into ions) completely.

Strong acids such as hydrochloric acid, nitric acid and sulfuric acid completely ionise in water.

The pH of a solution is related to the concentration of hydrogen ions; the higher the concentration of hydrogen ions the lower the pH. So, an aqueous solution of a strong acid (such as hydrochloric acid) will have a higher concentration of hydrogen ions and, therefore, a lower pH than an aqueous solution of a carboxylic acid of the same concentration.

? Test Yourself

1. What is the name of the functional group found in alcohols?

2. What is the molecular formula of propanol?

3. How can carboxylic acids be identified?

4. What is the general formula of alcohols?

★ Stretch Yourself

1. What are the products of the complete combustion of the alcohol ethanol?

Analysis

Modern Methods of Analysis

Compared to the more traditional laboratory methods, modern instrumental methods of analysing chemicals are faster, more sensitive, more accurate and require smaller sample sizes.

Qualitative analysis allows scientists to identify which components are present; **quantitative** analysis allows them to decide how much of each component is present.

It is important that scientists test a sample that is representative of the whole material being tested and that any sample is collected, stored and prepared carefully to prevent contamination, which would invalidate the results.

Many analytical techniques use samples that have been dissolved to form solutions.

Chromatography

Chromatography is used to **separate** the components of a mixture. It is used to analyse colouring agents in foods, flavourings and drugs. The technique is used in the food industry and by forensic scientists.

A small spot of the substance being analysed is placed on a pencil line towards the bottom of the chromatography paper. The chromatography paper is then placed end down into a beaker containing a small amount of the solvent being used. The solvent moves up the paper, carrying the soluble components. When the solvent front nearly reaches the top of the paper the paper is removed and the solvent is allowed to evaporate:
- Aqueous solvents contain water and are useful for many ionic compounds.
- Non-aqueous solvents do not contain water.

The more soluble a component is in the moving solvent the further it will move up the chromatography paper.

Build Your Understanding

The Rf value is used to compare the distance the component has moved compared to the distance the solvent front has moved.

$$Rf = \frac{\text{distance moved by the component}}{\text{distance moved by the solvent front}}$$

The solvent front has moved 6.0 cm.

The yellow spot has moved 3.0 cm.

The Rf value for the yellow spot $= \dfrac{3.0 \text{ cm}}{6.0 \text{ cm}} = 0.50$

If the same solvent and the same conditions are used, the Rf value would be the same for a given component. If the Rf value for an unknown compound is determined it can be identified by comparing this value with the Rf values for known compounds. Unfortunately, similar compounds often have quite similar Rf values.

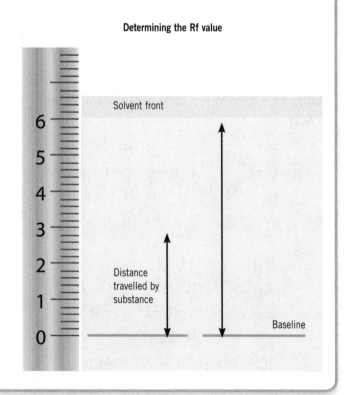

Determining the Rf value

Solvent front

Distance travelled by substance

Baseline

Spectroscopy

Infra-red spectroscopy is a technique used to identify the **bonds** present in organic compounds. This allows scientists to suggest the functional groups that are likely to be present. However, all the members of the same homologous series have the same functional group so it can be difficult to identify the exact compound present.

Atomic spectroscopy is used to identify the type and concentration of the atoms or ions present in a sample. For example, if a paint sample from a car was found at a crime scene, forensic scientists could use this technique to identify the make and age of the car involved.

Build Your Understanding

To identify the bonds present, scientists look for peaks at characteristic wave numbers in infra-red.

$C=O$ bonds absorb at around 1700 cm^{-1}, $O-H$ groups in alcohols are around $3200-3500$ cm^{-1} and $O-H$ groups in carboxylic acids are around $2500-3500$ cm^{-1}. The breathalyser used by police officers to measure the concentration of alcohol in the breath is based on this technique.

Mass Spectrometry

Mass spectrometry is used to find the accurate relative atomic or formula mass of a compound. Gas chromatography can be linked to mass spectrometry in a technique known as **GC-MS**. As the sample leaves the gas chromatogram, it is fed into the mass spectrum and the relative molecular mass of each substance can be identified.

The heaviest peak in the sample is known as the molecular ion and can be used to identify the molecular mass of the compound.

Gas Chromatography

Gas chromatography is used to identify organic compounds with low boiling points. The retention time is the time that it takes a component to pass through the column of the gas chromatogram.

Different components take different times to move through this column and so have different retention times. Unknown compounds can be identified by comparing their retention times with the retention times for known compounds. The areas under the peaks in the chromatogram are proportional to the amount of each compound in the sample. The number of peaks in the chromatogram shows the number of components in the sample.

Learn how gas chromatography and thin layer chromatography work.

Test Yourself

1. What are the advantages of modern methods of instrumental analysis over traditional methods?
2. Why is chromatography useful?
3. What is infra-red spectroscopy used for?
4. Why can't infra-red spectroscopy be used to identify the exact compound present?

Stretch Yourself

1. In separating a sample of food colour containing a blue and a green component, the chromatogram produced showed the solvent front moved 8.0 cm while the green component moved 6.0 cm and the blue component moved 5.0 cm. Calculate the Rf values for the green and blue components.

Cosmetics

Carboxylic Acids

Carboxylic acids are a family of organic compounds with the functional group **carboxyl**, COOH. Carboxylic acids are weak acids that react with metals, alkalis and carbonates. They have rather unpleasant smells: they are responsible for the smell of sweaty training shoes and rancid butter. The well-known carboxylic acid ethanoic acid is found in **vinegar**, which can be used to flavour food or as a preservative.

Examples of carboxylic acids

Perfumes

Traditional perfumes contain plant and animal extracts, such as rose and musk.

Today, cheaper **synthetic** fragrances, including esters, are often used. These are alternatives to materials made from living things.

A good perfume should:

- Evaporate easily from the skin so that the particles can be smelt.
- Be **non-toxic** and not react with water (so it does not react with sweat).
- Not irritate the skin (so it does not damage the skin).
- Be **insoluble** in water (so it is not washed off easily).

Build Your Understanding

Perfumes evaporate because, although there are strong forces of attraction *within* perfume particles, there are weaker forces of attraction *between* perfume particles. When the perfume is put on the skin, some of the particles gain enough energy to evaporate. The perfume particles can then travel through the air and be smelt.

Esters

Esters are a family of organic compounds formed when alcohols react with carboxylic acids. Esters have pleasant fruity smells and flavours. Esters are also used as solvents and as **plasticisers**.

The ester ethyl ethanoate is formed when ethanol is reacted with ethanoic acid using an acid **catalyst**.

Esters are very useful chemicals. Esters are described as being **volatile** because they **evaporate** easily.

They are used as cheap alternatives to naturally occurring compounds in perfumes and body sprays, and as flavourings in foods such as yoghurts.

Esters can be used as flavourings

Animal Testing

Cosmetic products, such as perfumes, have to be tested before they can be sold. Using animals to test cosmetics is banned in the UK, but living animals are used to test other products such as new medicines:

- Some people are against animal testing. They believe this causes avoidable suffering to animals.
- Other people think that animal testing is the best way to ensure that products are safe to use and that it allows us to develop new medicines that can save lives.

Making Solutions

Many people spend a lot of money on cosmetic items such as perfumes and make up. A solution is made from a solvent and a solute:

- The liquid, which does the dissolving, is called the solvent.
- The solid, which is dissolved, is called the **solute**.
- The **solution** is the mixture of the solvent and the solute. Solutions do not separate into layers (emulsions do).
- Water is a good solvent for many solids, but not everything dissolves in water; other liquids, like esters, can also be used.

The forces of attraction between particles decide whether or not something will dissolve.

Build Your Understanding

If a substance dissolves in a particular solvent, it is described as soluble. Nail varnish is soluble in nail varnish remover.

If a substance does not dissolve in a particular solvent, it is described as insoluble. Nail varnish is insoluble in water.

There are strong forces of attraction between water molecules. There are also strong forces of attraction between nail varnish particles. In fact, there are three lots of forces:

- The forces of attraction between water molecules.
- The forces of attraction between nail varnish particles.
- The forces of attraction between water molecules and nail varnish particles.

The first two forces are stronger than the last: nail varnish particles do not intermingle with the water molecules and nail varnish does not dissolve when you wash your hands in water. However, nail varnish can be removed using other solvents.

Nail varnish is soluble in nail varnish remover

Organic Chemistry and Analysis

❓ Test Yourself

1. What is the liquid that dissolves something called?
2. What is the solid that is dissolved called?
3. What term is used to describe the mixture made when a solid dissolves in a liquid?
4. Why are synthetic particles used in some perfumes?

⭐ Stretch Yourself

1. Describe how aftershave can be smelt from the other side of the room in terms of the particles involved and their movement.
2. Why must perfumes evaporate easily?

Complete these exam-style questions to test your understanding. Check your answers on page 122. You may wish to answer these questions on a separate piece of paper.

1 A student made a salad dressing from vinegar, olive oil and herbs. All the ingredients were placed in a glass bottle and then shaken together.

a) What do we call mixtures of substances, such as oil and vinegar, that do not normally mix together? (1)

b) Many shop-bought salad dressings do not need to be shaken before use. What is the name given to chemicals that produce stable mixtures of oil and vinegar, and how do these chemicals work? (2)

2 Crude oil is extracted from the Earth's crust. It is a mixture of many substances; the most important ones are hydrocarbons. Propane is obtained from crude oil. It can be used as a fuel. This diagram represents one molecule of propane.

```
    H   H   H
    |   |   |
H - C - C - C - H
    |   |   |
    H   H   H
```

a) Give the molecular formula of propane. (1)

b) Is propane a hydrocarbon? Explain your answer. (1)

c) The components of crude oil can be separated using fractional distillation. The large hydrocarbon molecules separated during fractional distillation are not very useful. These large hydrocarbon molecules can be broken down into smaller, more useful hydrocarbons by cracking. Cracking is an example of a thermal decomposition reaction. Explain why this reaction can be described as an example of thermal decomposition. (1)

d) A decane, $C_{10}H_{22}$, molecule can be split into two smaller molecules by cracking.

   i) Complete the equation to show this reaction.

   $C_{10}H_{22} \rightarrow$ _____ $+ C_2H_4$ (1)

   ii) Name the product of the cracking of decane that has the formula C_2H_4. (1)

   iii) What is C_2H_4 used to make? (1)

3 Carbon atoms form four bonds with other atoms. This means that carbon atoms can be made into an enormous number of compounds. The molecules below all contain atoms of the element carbon.

Molecule A

$$H-\underset{\underset{H}{|}}{\overset{\overset{H}{|}}{C}}-\underset{\underset{H}{|}}{\overset{\overset{H}{|}}{C}}-O-H$$

Molecule B

$$H-\underset{\underset{H}{|}}{\overset{\overset{H}{|}}{C}}-\underset{\underset{H}{|}}{\overset{\overset{H}{|}}{C}}-\underset{\underset{H}{|}}{\overset{\overset{H}{|}}{C}}-H$$

Molecule C

$$H-\underset{\underset{H}{|}}{\overset{\overset{H}{|}}{C}}-\underset{\underset{H}{|}}{\overset{\overset{H}{|}}{C}}=\overset{\overset{H}{|}}{C}-H$$

Molecule D

$$H-\underset{\underset{H}{|}}{\overset{\overset{H}{|}}{C}}-\underset{\underset{H}{|}}{\overset{\overset{H}{|}}{C}}-\underset{\underset{H}{|}}{\overset{\overset{H}{|}}{C}}=\overset{\overset{H}{|}}{C}$$

a) Is molecule A a hydrocarbon? Explain your answer. (1)

....

b) Is molecule B saturated? Explain your answer. (1)

....

c) State the name of molecule B. (1)

....

d) To which family of organic compounds does molecule C belong? (1)

....

e) State the name of molecule C. (1)

....

f) To which family of organic compounds does molecule D belong? (1)

....

g) State the name of molecule D. (1)

....

4 New products, such as medicines, must be tested on animals before they can be sold. Some people do not agree that animals should be involved in the testing process. Use your knowledge and understanding of the topic to outline the arguments for and against testing new products on animals.

You should make sure your answers are written using good spelling, punctuation and grammar. (6)

....

....

....

....

....

....

How well did you do?

| 0–6 | Try again | 7–12 | Getting there | 13–17 | Good work | 18–22 | Excellent! |

Metals

Metallic Structure

Metals have a giant structure. In metals, the electrons in the highest energy shells (outer electrons) are not bound to one atom but are **delocalised**, or free to move through the whole structure. This means that metals consist of positive metal ions surrounded by a 'sea' of negative electrons. **Metallic bonding** is the attraction between these positive ions and the negative electrons. This is an **electrostatic** attraction.

Metallic structure

Moving electrons can carry the electric charge or thermal (heat) energy.

✓ Maximise Your Marks

Remember, the metallic bond is the electrostatic attraction between the positive metal ions and the delocalised electrons.

Remember to say that metals conduct electricity because the delocalised electrons can move. Do not talk about atoms or ions moving.

Properties of Metals

Metallic bonding means that metals have several very useful properties:

- The free electrons mean that metals are **good electrical conductors**.
- The free electrons also mean that metals are **good thermal conductors**.
- Metals can be drawn into wires as the ions slide over each other.
- Metals can also be hammered into shape (they are malleable).
- Most metals have **high melting points** because lots of energy is needed to overcome the strong metallic bonds.

Non-metals are found on the right-hand side of the periodic table. They tend to be **poor electrical and thermal conductors**. Non-metals generally have **low melting points** and boiling points and are sometimes gases at room temperature.

Smart Alloys

Smart alloys are new materials with amazing properties.

One famous example of a smart alloy is **nitinol**. Nitinol is an alloy of nickel and titanium.

Some smart alloys have a **shape memory**. When a force is applied to a smart alloy it **stretches** or bends. When a smart alloy is heated up, however, it returns to its original shape.

Smart alloys appear to have a shape memory because they are able to exist in two solid forms. A temperature change of 10–20 degrees is enough to cause smart alloys to change forms.

Shape-memory polymers behave in a similar way, returning to their original shape when heated.

Smart Alloys (cont.)

At low temperatures, smart alloys exist in their low temperature form.

Low temperature form

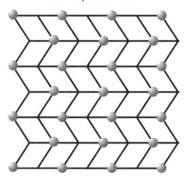

If a force is applied to the alloy it can be distorted to the low temperature, deformed form of the alloy.

Low temperature, deformed form of the alloy

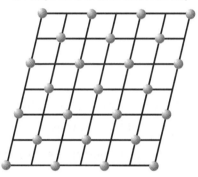

When the alloy is heated, it changes to the higher temperature form.

Higher temperature form

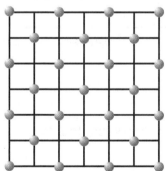

In shape-memory alloys, the low temperature form and the high temperature form are the same shape and size, so when they are heated smart alloys appear to have a shape memory. Nitinol is used in some dental braces.

Shape-memory alloys are also used in spectacle frames and as stents in damaged blood vessels.

Superconductors and Electrical Resistance

Some metals can behave as **superconductors** at very low temperatures. Metals can conduct electricity because they have delocalised electrons that can move. Metals normally have a **resistance** to the current that is flowing. Energy is lost as the current overcomes this resistance and the metal warms up.

Build Your Understanding

Superconductors are special because they have little or no resistance. The advantages of using superconductors include:

- If there is no resistance then no energy is lost when a current flows.
- As the resistance decreases the current can flow faster, so super-fast circuits can be developed.
- They can be used to make powerful electromagnets.

Despite these advantages, superconductors are not widely used because they only work below a critical temperature. Although this varies for different superconductors, the current critical temperatures are around −170 °C; until they work at room temperature, their use is likely to be limited.

? Test Yourself

1 What is metallic bonding?

2 Why are metals good electrical conductors?

3 Which metals are used to make nitinol?

4 Why do metals have high melting points?

★ Stretch Yourself

1 Give a use of nitinol.

2 Why is the use of superconductors limited at present?

Group 1

The Alkali Metals

The **elements** in group 1, on the far left-hand side of the periodic table, are known as the **alkali metals**. They are soft metals with **low melting points**.

They react with water to form hydroxide solutions. They react with halogens to form salts.

Examples of alkali metals are lithium (Li), sodium (Na) and francium (Fr).

Build Your Understanding

Rubidium and caesium belong to group 1, but are too **reactive** for use in schools. Further down the group alkali metals get more reactive; they react more vigorously with water. Alkali metals are so reactive that they must be stored under oil to prevent them reacting with moisture or oxygen. Alkali metals are shiny when freshly cut, but they tarnish quickly as they react with oxygen. Gloves and goggles should be worn when using alkali metals.

Reaction with Water

Alkali metals have **low densities**; lithium, sodium and potassium are all less dense than water. The alkali metals become denser down the group. When alkali metal atoms react they lose the single electron in their outermost shell to form ionic compounds in which the alkali metal ions have a 1+ charge.

For example:

$$Na \rightarrow Na^+ + e^-$$

The alkali metals react with water to form strongly **alkaline hydroxide** solutions and hydrogen gas:

Metal + Water → Metal Hydroxide + Hydrogen

Metal	Observations When Metal Reacts with Water	Equation
Lithium, Li	Metal floats on water. Some bubbles seen.	Lithium + Water → Lithium Hydroxide + Hydrogen
Sodium, Na	Metal forms a molten ball that moves around on the surface of the water. Many bubbles seen.	Sodium + Water → Sodium Hydroxide + Hydrogen
Potassium, K	The metal reacts even more vigorously than sodium (it can ignite). Lots of bubbles are seen and the hydrogen formed burns with a lilac flame.	Potassium + Water → Potassium Hydroxide + Hydrogen

✓ Maximise Your Marks

To get a top grade, you need to be able to write the symbol equations to sum up these three reactions:

$$2Li + 2H_2O \rightarrow 2LiOH + H_2$$

$$2Na + 2H_2O \rightarrow 2NaOH + H_2$$

$$2K + 2H_2O \rightarrow 2KOH + H_2$$

Why Group 1 Metals All React in a Similar Way

Alkali metals have just one electron in their outer shell and so have similar properties because they have similar electron structures.

Alkali metals react with non-metals to form ionic compounds. For example, sodium reacts with chlorine to form sodium chloride:

Sodium + Chlorine → Sodium Chloride

When sodium is burned, it reacts with oxygen to form sodium oxide:

Sodium + Oxygen → Sodium Oxide

When they react, an alkali metal atom loses its outer electron to form an ion with a 1+ charge:

$Na \rightarrow Na^+ + e^-$

The alkali metal atom has lost an electron so it is **oxidised**.

Alkali metals form solid white ionic compounds that dissolve to form colourless solutions.

✓ Maximise Your Marks

To get a top grade, you need to be able to write the symbol equations to sum up these reactions:

$2Na + Cl_2 \rightarrow 2NaCl$

$4Na + O_2 \rightarrow 2Na_2O$

Build Your Understanding

Down the group the outermost electron is further from the nucleus.

Further down the group there are more shells shielding the outer electron from the atom's nucleus, so it is easier for atoms to lose their outer electron.

Why Melting and Boiling Points Decrease down the Group

Melting and boiling points decrease down the group.

Alkali metals are held together by metallic bonding. Metallic bonding is the attraction between the positive metal ions and the 'sea' of negative electrons.

Forces of Attraction

The atoms get larger down the group. The strength of the metallic bonding decreases because the forces of attraction become weaker further down the group so it takes less energy to overcome these forces and therefore the melting points and boiling points will decrease.

? Test Yourself

1. Name the first three metals in group 1.
2. How many electrons are present in the outer shell of all group 1 metals?
3. Why do all the group 1 metals have similar properties?
4. What type of compounds do group 1 metals form?

★ Stretch Yourself

1. Potassium reacts with chlorine to produce the compound potassium chloride.
   a) Write a word and symbol equation to sum up this reaction.
   b) Is the potassium oxidised or reduced in this reaction? Explain your answer.

Extraction of Iron

Methods of Extracting Metals

Metals are very useful materials but they are normally found combined with other elements in **compounds** in the Earth's crust. The more **reactive** a metal is, the harder it is to remove it from its compound.

Gold is so **unreactive** that it is found uncombined, but most other metals are found in compounds. Occasionally, rocks are found that contain metals in such high concentrations that it is economically worthwhile to extract the metal from the rock. Such rocks are called **ores**. The ores are **mined**. The metals can be **extracted** using **chemical reactions**. The exact method chosen depends on the reactivity of the metal and the purity of the metal required.

Extracting Iron

Iron is an **element**. Elements are substances that are made of only one type of atom. There are only about 100 different elements. Iron is an extremely important metal. It is extracted from iron ore in a **blast furnace**. Iron is less reactive than carbon and can be extracted from iron oxide by reducing the metal oxide with carbon.

The solid **raw materials** in the blast furnace are iron ore, coke (a source of the element carbon) and limestone (which reacts with impurities). A fourth raw material, the gas air, is also used. The main ore of iron is **haematite**. This ore contains the compound iron (III) oxide, Fe_2O_3.

Iron ore

Build Your Understanding

The balanced symbol equation for the displacement reaction of iron and copper sulfate is:

$$Fe + CuSO_4 \rightarrow FeSO_4 + Cu$$

The iron atoms are oxidised while the copper ions are reduced.

In a similar way, copper will displace silver from a solution of silver nitrate. The copper atoms are oxidised while the silver ions are reduced:

Copper + Silver Nitrate → Copper Nitrate + Silver

$$Cu + 2AgNO_3 \rightarrow Cu(NO_3)_2 + 2Ag$$

In competition reactions, when a more reactive metal is heated with the oxide of a less reactive metal, the more reactive metal removes the oxygen from the less reactive metal oxide. If iron is heated with copper (II) oxide the more reactive metal, iron, removes the oxygen from the less reactive metal, copper:

Iron + Copper (II) Oxide → Iron (II) Oxide + Copper

$$Fe + CuO \rightarrow FeO + Cu$$

Displacement Reactions

In **displacement** reactions, a more reactive metal takes the place of a less reactive metal.

Iron is more reactive than copper, so if an iron nail is placed in a solution of blue copper (II) sulfate the nail changes colour from silver to pink–orange and the blue solution fades. There is also a slight temperature rise.

The more reactive metal, iron, displaces the less reactive metal, copper, from a solution of its compound:

Iron + Copper (II) Sulfate → Iron (II) Sulfate + Copper

What Happens in the Blast Furnace?

- Hot air is blasted into the furnace. The oxygen in the air reacts with the carbon in the coke to form carbon dioxide and to release energy:

Carbon + Oxygen → Carbon Dioxide

$$C(s) + O_2(g) → CO_2(g)$$

- At the very high temperatures inside the blast furnace, carbon dioxide reacts with more carbon to form carbon monoxide:

Carbon Dioxide + Carbon → Carbon Monoxide

$$CO_2(g) + C(s) → 2CO(g)$$

- The carbon monoxide reacts with iron oxide to form iron and carbon dioxide:

Carbon + Iron (III) → Iron + Carbon
Monoxide   Oxide   Dioxide

$$3CO(g) + Fe_2O_3(s) → 2Fe(l) + 3CO_2(g)$$

Haematite contains many impurities including substantial amounts of **silicon dioxide (silica)**. Limestone is added to the blast furnace because it reacts with these silica impurities to form **slag**. Slag has a low density and floats on top of the molten iron ore where it can be removed. Slag is used in road building and in the manufacture of fertilisers.

Blast furnace

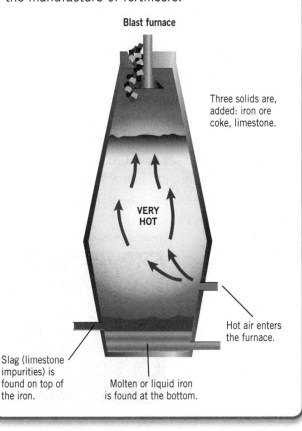

Three solids are, added: iron ore coke, limestone.

VERY HOT

Hot air enters the furnace.

Slag (limestone impurities) is found on top of the iron.

Molten or liquid iron is found at the bottom.

Oxidation and Reduction

In many examples, oxygen is gained in **oxidation** reactions and oxygen is lost in **reduction** reactions. Oxidation and reduction always happen together. During the extraction of iron the carbon is first oxidised to carbon monoxide. The iron oxide is reduced to iron and the carbon monoxide is oxidised to carbon dioxide.

Build Your Understanding

The high temperatures in the blast furnace means the iron that is made is a liquid. This molten iron is dense and sinks to the bottom of the furnace where it can be removed. Iron ore is mainly reduced by the gas carbon monoxide but some is reduced by carbon.

Corrosion is an example of an oxidation reaction. The less reactive a metal is the more slowly it corrodes.

Copper can be extracted from copper oxide by heating it with carbon:

Copper Oxide + Carbon → Copper + Carbon
Dioxide

❓ Test Yourself

1. What happens in a displacement reaction?
2. What colour is copper (II) sulfate solution?
3. Give the word equation for the reaction between copper (II) sulfate and iron.

⭐ Stretch Yourself

1. Zinc is more reactive than copper. Write the word and symbol equation to sum up the reaction between zinc metal and copper (II) sulfate solution. Explain which metal is oxidised and which metal is reduced in this reaction.

Metals and Tests

Iron and Steel

Preventing Iron from Rusting

Iron **corrodes**, or **rusts**, faster than most other **transition metals**. Rusting involves **oxidation**. It requires the presence of both oxygen and water and produces hydrated iron (III) oxide.

Rusting is accelerated by salt water and by acid rain. Iron can be prevented from rusting by completely removing it from contact with either oxygen or water.

Iron + Oxygen + Water → Hydrated Iron (III) Oxide

Oxidation and reduction reactions always involve electrons. In oxidation reactions electrons are always lost. In reduction reactions electrons are always gained. In this example, the iron is oxidised and oxygen is reduced. The iron is the reducing agent and the oxygen is the oxidising agent.

Coating the Iron

Painting or **coating** iron in plastic, oil or with tin plate can stop oxygen and water from reaching the metal. If the coating is damaged, however, the iron will start to rust.

Tin is less reactive than iron. If the tin is scratched, the iron will react by losing electrons, even faster than it would normally do, so the iron will rust more quickly.

Sacrificial Protection

Sacrificial protection involves placing iron in contact with a more reactive metal like zinc or magnesium to prevent rusting. The iron is protected because the more reactive metal reacts by losing electrons, instead of the iron – which is why the method is called sacrificial protection.

Galvanising protects iron by coating it in a layer of zinc, another sacrificial metal. The zinc layer stops oxygen and water from reaching the iron.

Alloying the Metal

Alloys are mixtures containing one or more metal. They are made by mixing molten metals together. Most **pure metals**, such as copper and aluminium, are too soft for many uses so molten mixtures of similar metals are combined to form alloys.

Pure metals are layered because all the atoms are the same size and these atoms can slide over each other.

In alloys the atoms are different sizes; this causes disruption of the layers so they cannot pass over each other.

Alloys such as brass, bronze, steel, solder and amalgam are harder and more useful than the pure metals.

Cast Iron

The iron that is made in a blast furnace contains about 4 per cent of the element carbon. If this iron is allowed to cool down and solidify, it forms **cast iron**. Cast iron contains about 96 per cent pure iron and is used to make objects like drain covers. It is hard, strong and does not rust.

Cast iron does have one notable disadvantage: it is brittle and can crack easily.

Drain covers are made from cast iron

Wrought Iron

Wrought iron is made by removing the impurities from cast iron.

It is much softer than cast iron and is used to make objects like gates.

Build Your Understanding

Wrought iron is softer and easier to shape than cast iron because of its structure. It is made from almost pure iron. The iron atoms form very regular layers that are able to slip over each other easily.

Wrought iron structure

Steel

Most of the iron made in the blast furnace is used to produce steel. Mixing iron with other metals and carbon to form alloys, such as stainless steel, will also protect the metal. To make steel:

- Any carbon impurities must first be removed from the iron, to produce pure iron.
- Other metals and carefully controlled amounts of the non-metal element carbon are added to the iron.

Steel is much harder than wrought iron because it consists of atoms of different elements. These atoms are different sizes and cannot pack together to form a regular structure. This irregular structure makes it very difficult for layers of atoms to slide over each other, which makes the steel very hard.

Steel structure

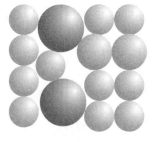

By carefully controlling the amount of carbon that is added to steel, scientists can produce a metal that has exactly the right properties for each particular job:

- **Low carbon steels** are soft and easy to shape. Objects such as car bodies are made from low carbon steels.
- **Medium carbon steels** are harder, stronger and less easy to shape. Objects such as hammers are made from medium carbon steels.
- **High carbon steels** are hard, strong, brittle and hard to shape. Objects such as razor blades are made from high carbon steels.

Stainless steel is a very widely used alloy. It consists of 70 per cent iron, 20 per cent chromium and 10 per cent nickel. Stainless steel is extremely resistant to corrosion.

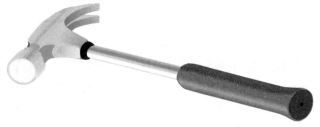

Hammers are made from medium carbon steel

? Test Yourself

1. What must be present for iron to rust?
2. Give the chemical name for rust.
3. What are alloys?

★ Stretch Yourself

1. A sample of pure silver was found to be too soft for making jewellery. Explain how the silver could be made more hard-wearing in terms of the atoms involved.

Aluminium

Bauxite

Aluminium is abundant in the Earth's crust, but it is also very **reactive** and so it is hard to **extract** from its **ores**. Consequently, aluminium is more expensive than iron.

The main ore of aluminium is called **bauxite**. Bauxite contains the compound aluminium oxide, Al_2O_3.

Unfortunately, bauxite is often found in environmentally sensitive areas such as the Amazon rainforests.

Extracting bauxite from these places brings jobs and money to the area, but can also scar the landscape and harm local wildlife. New roads can damage the areas around the mines and local people can be displaced by the development.

One way to protect the areas where bauxite is found is for people to simply **recycle** their old aluminium cans.

Recycling has many advantages:
- Less bauxite will need to be extracted.
- Landfill sites will not be filled up with discarded aluminium cans.
- Much less energy is used by recycling than extracting aluminium straight from its ore.

Properties of Aluminium

Although pure aluminium is quite **soft**, when it is alloyed with other metals it becomes much harder and stronger.

Aluminium alloys combine high strength with low density. This makes aluminium a very useful metal for products like aeroplanes and mountain bikes.

Aluminium is also a good electrical conductor.

Uses of Aluminium

Aluminium is a reactive metal and yet it is widely used to make drinks cans. Aluminium is much less reactive than its position in the reactivity series would suggest. This is because aluminium quickly reacts with oxygen to form a thin layer of **aluminium oxide**. This layer stops the aluminium metal from coming into contact with other chemicals and so prevents any further reaction. The layer of aluminium oxide means that it is quite safe to drink fizzy, acidic drinks from aluminium cans.

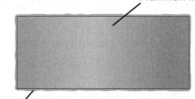

Aluminium reacts with oxygen to form aluminium oxide

Aluminium metal

Layer of aluminium oxide

The layer of aluminium oxide stops aluminium from reacting further.

The Extraction of Aluminium

Aluminium is more reactive than carbon and so it is extracted using **electrolysis**, even though this is a very expensive method. For electrolysis to occur, the aluminium ions and oxide ions in bauxite must be able to move. This means that the bauxite has to be either heated until it melts or dissolved in something.

Build Your Understanding

Bauxite has a very high melting point and heating the ore to this temperature is very expensive. Fortunately, another ore of aluminium, called cryolite, has a much lower melting point. First, the cryolite is heated up until it melts and then the bauxite is dissolved in the molten cryolite. Extracting aluminium from its ore requires a lot more energy than extracting iron from its ore.

Electrolysis

Aluminium can be extracted by electrolysis.

- By dissolving the aluminium oxide in cryolite, both the aluminium, Al^{3+}, ions and the oxide, O^{2-}, ions can move.
- During electrolysis, the aluminium, Al^{3+} ions are attracted to the negative electrode (the cathode) where they pick up electrons to form aluminium, Al, atoms. The aluminium metal collects at the bottom of the cell where it can be gathered:

  Aluminium ions + electrons → aluminium atoms

- The oxide, O^{2-}, ions are attracted to the positive electrode (the anode) where they lose electrons to form oxygen molecules:

  Oxide ions – electrons → oxygen molecules

- The oxygen that forms at the positive electrode readily reacts with the carbon, graphite, electrode to form carbon dioxide. The electrodes, therefore, must be replaced periodically. Extracting aluminium is expensive because lots of energy is required and because there are lots of stages in the process.

Electrolysis of bauxite (aluminium oxide)

Carbon lining as negative electrode
Positive carbon electrodes
+
−
Steel tank
Tap hole
Molten aluminium
Oxide ions
Purified aluminium oxide in molten cryolite
Aluminium ions

Oxidation and Reduction

In the electrolysis of aluminium oxide:

- Aluminium ions are reduced to aluminium atoms.
- Oxide ions are oxidised to oxygen molecules.

Reduction reactions happen when a substance gains electrons. In this case, each aluminium ion gains three electrons to form an aluminium atom.

Oxidation reactions occur when a species loses electrons. In this case, two oxide ions both lose two electrons to form an oxygen molecule.

Reduction and oxidation reactions must always occur together and so are sometimes referred to as **redox** reactions.

❓ Test Yourself

1. What is a mixture containing at least one metal called?
2. What is the formula of aluminium oxide?
3. What is the name of the method used to extract aluminium from its ore?

⭐ Stretch Yourself

1. In the extraction of aluminium, why is bauxite dissolved in molten cryolite?

Cars

Choosing Between Aluminium and Steel for Making Cars

Cars have become an indispensable part of many people's lives. However, cars do not last forever and their disposal can cause problems.

Cars are built using a wide range of different materials including:
- plastics
- glass
- copper
- aluminium
- steel
- fibres.

Steel is an alloy of iron. Steel and aluminium can both be used as construction materials. Steel and aluminium are both extracted from their ores, which are found in the Earth's crust. Although both these materials are metals they have different properties:
- Steel is magnetic; aluminium is non-magnetic.
- Steel is denser than aluminium.
- Steel corrodes readily; aluminium does not.

Conditions affect how quickly metals corrode. Acid rain, salt water and moist air all increase the rate of corrosion.

Like all metals, both steel and aluminium are good **electrical** and **thermal conductors**. They are both **malleable** and can be hammered into shape; this is particularly useful for making cars.

Car bodies have traditionally been made from steel. However, some cars are now being built from aluminium.

New laws will soon mean that a certain minimum amount of any new car must be recyclable.

✓ Maximise Your Marks

Learn the information in this table.

Advantages of Building Cars From Steel	Advantages of Building Cars From Aluminium
Steel is cheaper than aluminium, so steel cars are cheaper to buy	Aluminium cars do not corrode so aluminium cars will last longer
	Aluminium is less dense than steel so aluminium cars will be lighter than steel cars and will have a better fuel economy

Hydrogen as a Fuel

Hydrogen can be used as a **fuel**. When it burns it reacts with oxygen to form water. The reaction is **exothermic** (it releases heat energy). The hydrogen burns with a clean blue flame.

Hydrogen is highly flammable so must be stored safely.

Hydrogen can be used as a fuel for cars, aeroplanes and boats. Hydrogen does not occur naturally and has to be manufactured.

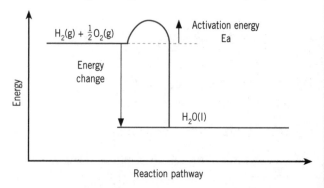

The energy level diagram for the combustion of hydrogen

Fuel Cells

Fuel cells are a very efficient way of producing electrical energy. Most fuel cells use hydrogen, but other fuels, such as ethanol, can be used. Typically, hydrogen and oxygen are fed into the fuel cell and chemical reactions inside the cell produce electricity.

Traditionally, **fossil fuels** like petrol have been used as a fuel for cars. However, these fuels have many disadvantages:
- They are **non-renewable**.
- When they are burned they release carbon dioxide, which contributes to the **greenhouse effect**.

Fuel cells could also be used to power cars. If hydrogen was used, the cars would not produce carbon dioxide and, as the hydrogen is made by the decomposition of water, an effectively limitless amount of hydrogen gas would be available.

Hydrogen and oxygen can both be made from the electrolysis of water, H_2O. Twice as much hydrogen as oxygen is produced.

However, the use of **fuel cells** can still cause pollution. Many fuel cells contain poisonous **catalysts** that must be removed from the cell before it is disposed of. In addition, although the fuel cell itself does not produce carbon dioxide, water is currently decomposed to form hydrogen and oxygen using electricity that has been generated by burning fossil fuels – a process that does produce carbon dioxide.

Build Your Understanding

Fuel cells are ideal for remote situations where space is limited; they have been used to power spacecraft. Hydrogen fuel cells produce water, which is non-polluting and is essential for the astronauts. The fuel cells are also lightweight and small. As they have no moving parts, fuel cells are unlikely to break down and are very efficient. Traditional methods of producing electricity, for example in coal-fired power stations, involve many more stages. This makes them less efficient; older methods also produce more pollution.

In a typical fuel cell, a fuel such as hydrogen reacts with oxygen to create a potential difference.

Hydrogen is added at the anode.

The hydrogen molecules lose electrons to form hydrogen ions:

$$2H_2 \rightarrow 4H^+ + 4e$$

Oxygen is reduced at the cathode.

The hydrogen ions produced at the anode move through the electrolyte (the substance broken down) to the cathode where they join with oxygen and electrons to form water:

$$O_2 + 4H^+ + 4e^- \rightarrow 2H_2O$$

Overall:

Hydrogen + Oxygen $\rightarrow$ Water

$$2H_2 + O_2 \rightarrow 2H_2O$$

✓ Maximise Your Marks

A fuel cell produces electrical energy.

❓ Test Yourself

1. Which factors increase the rate at which steel rusts?

2. Why do cars made from aluminium last longer than cars made from steel?

3. Name the product made when hydrogen is burned in oxygen.

4. Why is hydrogen considered to be a non-polluting fuel?

⭐ Stretch Yourself

1. Write the word and symbol equation for the overall reaction inside a typical fuel cell.

Transition Metals

Properties of Transition Metals

Transition metals are found in the middle section of the periodic table. Copper, iron and nickel are examples of very useful transition metals. All transition metals have characteristic properties:

- High **melting points** (except for mercury, which is a liquid at room temperature).
- A high **density**.

They form **coloured compounds**:

- Copper compounds are blue or green.
- Iron (II) compounds are green.
- Iron (III) compounds are a 'foxy' red.

Reactions of Transition Metals

Transition metals are strong, tough, **good thermal and electrical conductors**, malleable and hard wearing. All transition metals are much less reactive than group 1 metals. They react much less vigorously with oxygen and water. Many transition metals can form ions with different charges. This makes transition metals useful catalysts for many reactions.

Transition metals are hard wearing

Titanium

Despite being very abundant in the Earth's crust, **titanium** is an expensive metal. This is because it is difficult to extract titanium from its ore. The main ore of titanium is **rutile**. Rutile contains the compound titanium oxide, TiO_2. Titanium is more reactive than carbon and so cannot be extracted simply by heating titanium oxide with carbon.

Build Your Understanding

The extraction of titanium is a complicated process:

- First, the titanium oxide is converted to titanium chloride.
- Next, the titanium chloride is reacted with molten magnesium. Magnesium is more reactive than titanium and a chemical reaction takes place in which titanium is displaced:

Titanium + Magnesium → Titanium + Magnesium Chloride   Chloride

The extraction of titanium involves many steps and requires a lot of energy, so it is very expensive.

Properties of Titanium

Titanium has some very special properties:

- It is very strong and hard when alloyed with other metals.
- It has a relatively low density.
- It is easy to shape.
- It has a very high melting point.
- It is very resistant to corrosion.

Titanium appears to be unreactive because the surface of titanium objects quickly reacts with oxygen to form a layer of titanium oxide. This layer prevents any further reaction taking place.

Titanium's properties mean that this metal is very useful. Titanium alloys are used to make replacement hip and elbow joints, aircraft and rockets and missiles.

Other Useful Metals

Copper has some very special properties:
- It is a good thermal and electrical conductor.
- It is easy to shape.
- It is very **unreactive** – even with water.
- It has an attractive colour and lustre.
- It is very resistant to **corrosion**.

Copper's properties mean that it is a very useful metal. Copper is used to make water pipes and tanks, saucepans and electrical wires.

Copper has many uses

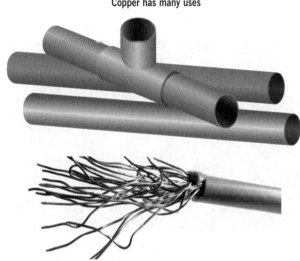

Iron made in the **blast furnace** is **strong** but **brittle**. Iron is often made into steel. **Steel** is strong and cheap and is used in vast quantities, but it is also heavy and may rust.

Iron and steel are useful structural materials. They are used to make buildings, bridges, ships, cars and trains.

Iron is used as a **catalyst** in the **Haber process**.

Gold is used to make jewellery and electrical components. Gold is a highly valued metal that has an attractive colour and lustre. It is also a good thermal conductor and, because of its low reactivity, is very resistant to corrosion. Pure gold is too soft for many uses so it is usually mixed with other metals to form alloys.

The **carat scale** and the **fineness scale** are both used to show the amount of pure gold in the alloy; in both cases, the higher the number the greater the proportion of gold.

Metal Alloys

Common alloys include:
- Amalgams, which contain mainly mercury and are used in tooth fillings.
- Brass, which is made from copper and zinc and are used to make musical instruments and coins.
- Bronze, which is made from copper and tin.
- Solder, which is made from lead and tin and is used to join electrical wires.
- Steel which is mainly iron.

Nickel is used as a **catalyst** in the manufacture of margarine.

❓ Test Yourself

1. In which section of the periodic table are the transition metals found?
2. Why is copper used for electrical wiring and for water pipes?
3. In which process is iron used as a catalyst?
4. How can you tell how pure a sample of gold is?

⭐ Stretch Yourself

1. Why are many transition metal compounds useful catalysts?
2. Give the word equation for the displacement reaction between titanium chloride and magnesium.
3. Why is the extraction of titanium very expensive?

Copper

Extraction of Copper

Copper is an unreactive metal and there are several copper ores. Copper has been known since ancient times, so the richest supplies of ores have been exhausted. Copper is now extracted from rocks that do not contain large amounts of the metal. This means that a lot of rock has to be **quarried** in order to extract enough copper and this can cause significant damage to the local area.

Copper is sometimes found uncombined (or native) in nature. The mineral malachite contains copper carbonate, $CuCO_3$. When it is heated, the copper carbonate breaks down to form copper oxide and carbon dioxide:

Copper Carbonate → Copper Oxide + Carbon Dioxide

$$CuCO_3 \rightarrow CuO + CO_2$$

The copper oxide produced reacts with carbon to form copper and carbon dioxide:

Copper Oxide + Carbon → Copper + Carbon Dioxide

$$2CuO + C \rightarrow 2Cu + CO_2$$

Malachite contains copper carbonate

Extracting Copper from Low-grade Ores

Scientists are developing ways to exploit copper from **low-grade ores**, which contain copper at lower concentrations than would normally be economically worthwhile to use. The idea is to **leach** copper out of the ores to form a solution, then to extract the copper from the solution using electrolysis or by displacement with scrap iron.

Metals can be extracted from low-grade ores using plants. As the plant grows, it takes up the metal, which accumulates in the plant's biomass. When the plant is harvested the **biomass** can be burned to produce a bio-ore. This process, called **phytomining**, allows scientists to exploit ores that had previously been uneconomic to use.

Bioleaching is the process of extracting metals from their ores using **bacteria**.

Advantages of bioleaching include:
- It is a simpler and cheaper process compared to traditional smelting methods.
- It causes less damage to the landscape than traditional methods.

Disadvantages of bioleaching include:
- It is a very slow process.
- There is a risk of pollution if toxic chemicals are allowed to escape into the environment.

Phytomining allows scientists to recover toxic metals from waste dumps and to reclaim **contaminated** areas of land. The recovered metals, such as nickel and cobalt, can be used for new purposes. The land can also be converted to new uses, such as agriculture or for building new homes.

Recycling copper is better for the environment than more extraction because fewer raw materials are needed. As less energy is required, the copper is cheaper to buy.

However, people have to be persuaded to recycle their waste metals rather than putting them into landfill sites. Sorting waste metals can also be very labour intensive and expensive.

Purification of Copper

Copper must be purified before it can be used for some applications, such as high-specification wiring.

Copper is purified using electrolysis:

- During the electrolysis of copper, impure copper metal is used as the positive electrode where copper atoms give up electrons to form copper ions.
- As the positive electrode dissolves away, any impurities fall to the bottom of the cell to form a sludge.
- Copper ions in the solution are attracted towards the negative electrode where the copper ions gain electrons to form copper atoms.
- The positive electrode gets smaller while the negative electrode gets bigger. In addition, the negative electrode is covered in very pure copper.

Electrolysis of copper

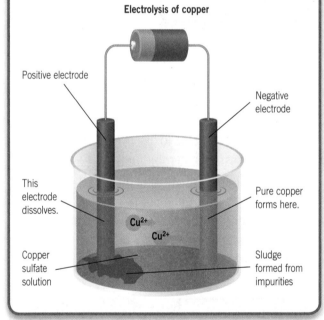

Positive electrode

Negative electrode

This electrode dissolves.

Pure copper forms here.

Copper sulfate solution

Sludge formed from impurities

Cu^{2+}

Cu^{2+}

Build Your Understanding

The following reactions take place during the electrolysis of copper. The reaction at the positive electrode:

Copper atoms − electrons → Copper ions

$$Cu \quad - \quad 2e^- \quad \rightarrow \quad Cu^{2+}$$

The reaction at the negative electrode:

Copper ions + electrons → Copper atoms

$$Cu^{2+} \quad + \quad 2e^- \quad \rightarrow \quad Cu$$

Copper Alloys

Pure copper is too soft for many uses. In pure copper the atoms are all the same size and so they form a regular arrangement. Copper is soft because the layers of atoms can slide easily over each other. Copper is often mixed with other metals to form **alloys**.

Bronze is made by mixing copper and tin. It is much harder than either copper or tin and consists of different sized atoms, so the atoms cannot pack together to form a regular structure. Bronze is hard because the layers of atoms cannot slide easily over each other. The invention of bronze was a major advance: it was used to make stronger tools and weapons.

The structure of bronze

❓ Test Yourself

1. Which method is used to purify copper?
2. If you wanted to coat a metal object with copper, which electrode should you make it?
3. Why can copper alloys be more useful than pure copper?

⭐ Stretch Yourself

1. Copper can be purified by electrolysis.
   a) Give the symbol equation for the reaction that takes place at the positive electrode during the purification of copper.
   b) Give the symbol equation for the reaction that takes place at the negative electrode during the purification of copper.

Chemical Tests 1

Gas Tests

Chemists use these tests to identify the following common gases.

Carbon dioxide: The gas is bubbled through limewater. The limewater turns cloudy.

Carbonates react with acids to produce carbon dioxide.

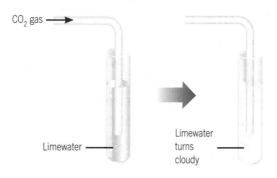

CO_2 gas →

Limewater

Limewater turns cloudy

Hydrogen: A lighted splint is placed nearby. The hydrogen burns with a squeaky pop.

'Squeaky pop'

H_2 gas

Chlorine: Place damp litmus paper in the gas. The litmus paper is bleached.

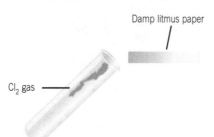

Damp litmus paper

Cl_2 gas

Oxygen: A glowing splint is placed in the gas. The splint relights.

Glowing splint

Cl_2 gas

Gas Tests (cont.)

Ammonia: Place damp red litmus paper in the gas. The damp red litmus paper turns blue.

NH_3 gas

Damp red litmus

✔ Maximise Your Marks

Make sure you are familiar with the tests for common gases – you will need to state the chemicals used and the results in the exam.

Flame Tests

Flame tests can be used to identify some metals present in salts. These elements give distinctive flame colours when heated because the light given out by a particular element gives a characteristic **line spectrum**. The technique of **spectroscopy** has been used by scientists to discover new elements, including caesium and rubidium:

- Clean a flame test wire by placing it into the hottest part of a blue Bunsen flame.
- Dip the end of the wire into water and then into the salt sample.
- Hold the salt in the hottest part of the flame and observe the colour seen.

Metal Ion Present	Colour in Flame Test
Lithium	Crimson
Sodium	Yellow/orange
Potassium	Lilac
Calcium	Red
Barium	Light green
Copper	Blue/green

✔ Maximise Your Marks

For flame tests, give the name of the metal ion responsible for a colour, not the name of a whole compound.

Hazard Symbols

Hazard symbols are a very effective way of alerting people to the dangers associated with different chemicals.

Toxic These substances can kill. They can act when you swallow them, breathe them in or absorb them through your skin. *Example: chlorine gas.*		**Corrosive** These substances attack other materials and living tissue, including eyes and skin. *Example: concentrated sulfuric acid.*	
Oxidising These substances provide oxygen, which allows other substances to burn more fiercely. *Example: hydrogen peroxide.*		**Irritant** These substances are not corrosive but they can cause blistering of the skin. *Example: calcium chloride.*	
Harmful These substances are similar to toxic substances but they are less dangerous. *Example: lead oxide.*		**Explosive** These substances are explosive. *Example: urea nitrate.*	
Highly Flammable These substances will catch fire easily. They pose a serious fire risk. *Example: hydrogen.*			

> ✓ **Maximise Your Marks**
>
> Questions are often asked about hazard symbols. Make sure you can identify what each symbol shows and explain what it means.

Useful Techniques in Science

There are many useful techniques that can be used by chemists:

- **Dissolving** is used to form a solution from a solute and a suitable solvent.
- **Crystallisation** is used to produce solid crystals from a solution.
- **Filtration** is used to separate an insoluble solid from a mixture of solid and liquid.
- **Evaporation** is used to turn a liquid into a gas. If a solution is evaporated to dryness the mass of the solute can be found.
- **Drying**, in an oven or in a desiccator, is used to remove water from a sample.

? Test Yourself

1. What is the test for hydrogen gas?
2. Which scientific technique is used to separate an insoluble solid from a mixture?

⭐ Stretch Yourself

1. Describe how you would carry out a flame test.

Metals and Tests

79

Chemical Tests 2

Formula of Ionic Compounds

Metal Ions	Non-metal Ions
Sodium, Na^+	Oxide, O^{2-}
Magnesium, Mg^{2+}	Chloride, Cl^-
Calcium, Ca^{2+}	Bromide, Br^-
Potassium, K^+	Hydroxide, OH^-
Iron (II), Fe^{2+}	Nitrate, NO_3^-
Iron (III), Fe^{3+}	Carbonate, CO_3^{2-}
Copper (II), Cu^{2+}	Sulfate, SO_4^{2-}

The compound magnesium oxide contains magnesium, Mg^{2+}, and oxide, O^{2-}, ions.
For every one magnesium ion, one oxide ion is required.

The overall formula for the compound is MgO.

Testing for Sulfate Ions

To test for the presence of sulfate ions in solution:
- Add dilute hydrochloric acid.
- Then add barium chloride solution.

A white precipitate of barium sulfate shows that sulfate ions are present in the original solution:

Barium	+	Sodium	→	Barium	+	Sodium
Chloride		Sulfate		Sulfate		Chloride

Symbol equation:

$$BaCl_2(aq) + Na_2SO_4(aq) \rightarrow BaSO_4(s) + 2NaCl(aq)$$

Ionic equation:

$$Ba^{2+}(aq) + SO_4^{2-}(aq) \rightarrow BaSO_4(s)$$

Testing for Halide Ions

It is important that a test for a particular ion gives a result that is **unique** to that ion for a positive identification to be made. These tests are very important and are used to check for the presence of chemicals in blood and to check the purity of drinking water.

Identifying Halide Ions			
Halide Ion	Test	Results	Ionic Equations
Chloride, Cl^-	Add dilute nitric acid then silver nitrate solution	Chloride ions give a white precipitate of silver chloride	$Ag^+(aq) + Cl^-(aq) \rightarrow AgCl(s)$
Bromide, Br^-	Add dilute nitric acid then silver nitrate solution	Bromide ions give a cream precipitate of silver bromide	$Ag^+(aq) + Br^-(aq) \rightarrow AgBr(s)$
Iodide, I^-	Add dilute nitric acid then silver nitrate solution	Iodide ions give a yellow precipitate of silver iodide	$Ag^+(aq) + I^-(aq) \rightarrow AgI(s)$

Metals and Tests

Identifying Carbonates

Metal **carbonates** react with dilute hydrochloric acid to form a salt, water and carbon dioxide gas.

To prove the gas produced is carbon dioxide, place a drop of limewater (calcium hydroxide $Ca(OH)_2$ solution) on a glass rod. If carbon dioxide is present the limewater turns cloudy:

Copper + **Hydrochloric** → **Copper** + **Water** + **Carbon**
Carbonate   **Acid**   **Chloride**   **Dioxide**

Symbol equation:

$$CuCO_3(s) + 2HCl(aq) \rightarrow CuCl_2(aq) + H_2O(l) + CO_2(g)$$

Ionic equation:

$$CuCO_3(s) + 2H^+(aq) \rightarrow Cu^{2+}(aq) + H_2O(l) + CO_2(g)$$

When copper carbonate is heated it decomposes to form copper oxide and carbon dioxide. This can be identified by a distinctive colour change: copper carbonate is green and copper oxide is black.

Testing for carbon dioxide

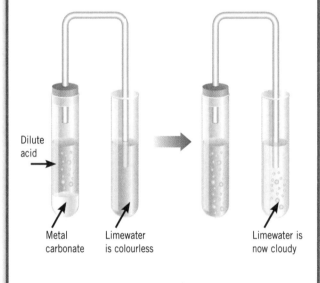

Dilute acid

Metal carbonate

Limewater is colourless

Limewater is now cloudy

Build Your Understanding

Dilute sodium hydroxide solution can be used to test for the presence of some transition metal ions in solution. The sodium hydroxide solution is added dropwise to the solution of the transition metal compound. The colour of the precipitate formed is used to identify the transition metal ion.

Transition Metal Ion	Results
Copper (II), Cu^{2+}	Blue precipitate of copper (II) hydroxide, $Cu(OH)_2$
Iron (II), Fe^{2+}	Green precipitate of iron (II) hydroxide, $Fe(OH)_2$ This turns brown as the Fe^{2+} ions are oxidised to Fe^{3+} ions

Sodium hydroxide can also be used to identify aluminium and calcium ions in solutions.

Metal Ion	Results
Aluminium, Al^{3+}	White precipitate that *does* dissolve in excess sodium hydroxide to form a colourless solution
Calcium, Ca^{2+}	White precipitate that *does not* dissolve in excess sodium hydroxide

To test for the presence of ammonium ions in a compound, add sodium hydroxide solution and then warm the resulting mixture. Then test for the presence of ammonia gas.

? Test Yourself

1. What is the test for the gas carbon dioxide?
2. What colour is the precipitate formed when copper (II) ions react with hydroxide ions?
3. What colour is the precipitate formed when iron (II) ions react with hydroxide ions?

★ Stretch Yourself

1. Give the ionic equation for the reaction between copper carbonate and hydrochloric acid.

Practice Questions

Complete these exam-style questions to test your understanding. Check your answers on page 123. You may wish to answer these questions on a separate piece of paper.

1 A blast furnace is used to extract iron from iron ore.

   a) Three solid raw materials are added to the blast furnace. One of the solid raw materials is iron ore. Name the other two.   (2)

   b) The main chemical compound in iron ore is iron (III) oxide. What is the main ore of iron called?   (1)

   c) What is the last raw material added to the blast furnace?   (1)

   d) Complete this equation to show the reaction between carbon monoxide and iron (III) oxide.   (1)

      $3CO + Fe_2O_3 \rightarrow 2Fe + 3\underline{\hspace{1em}}$

   e) In the blast furnace, iron is extracted from iron ore. Which of these words best describes what happens to the iron in iron oxide? Tick one box.   (1)

      ☐ Electrolysis   ☐ Oxidation
      ☐ Reduction   ☐ Neutralisation

   f) In the blast furnace, what does the limestone do?   (1)

2 a) What two substances must be present for iron to rust?   (1)

   b) Explain how coating an iron fence in plastic can stop the iron from rusting.   (1)

3 This diagram shows the structure of cast iron.

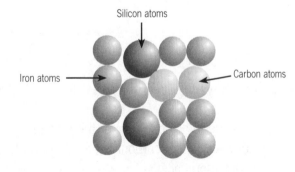

   a) Sketch a diagram to show the structure of wrought iron. Label the atoms in your diagram.   (2)

b) Would you expect cast iron or wrought iron to be harder? Explain your answer.   (1)

....

....

....

c) Steel can be used to make hammers. Which of these properties should a hammer have?
Tick one answer.   (1)

☐ Soft   ☐ Brittle
☐ Hard to shape   ☐ Strong

d) How does increasing the amount of carbon in steel affect how easy the carbon is to shape?   (1)

....

4 **a)** Aluminium can be extracted from aluminium oxide. What is the name given to this process?   (1)

....

b) Name the main ore of aluminium.   (1)

....

c) Name the second ore of aluminium that is also used in the extraction of aluminium.   (1)

....

d) Why is the second ore of aluminium used?   (1)

....

e) During the electrolysis of aluminium oxide, the aluminium ions move. Which electrode do
these ions move towards?   (1)

....

f) Which of these words best describes what happens to aluminium ions during electrolysis?
Tick one box.   (1)

☐ Displacement   ☐ Oxidation
☐ Reduction   ☐ Neutralisation

g) During the electrolysis of aluminium oxide, the oxide ions also move. Which electrode do
these ions move towards?   (1)

....

h) What material is the positive electrode made from?   (1)

....

How well did you do?

| 0–8 | Try again | 9–14 | Getting there | 15–18 | Good work | 19–22 | Excellent! |

Acids and Bases

Acids, Bases and Salts

Strong Acids

Acids and bases are chemical opposites. Some bases dissolve in water and are called alkalis.

Acidic solutions have a pH less than 7. Acidic compounds can be solids like citric acid and tartaric acid, liquids like sulfuric acid, nitric acid and ethanoic acid or gases like hydrogen chloride.

Some acids are described as strong. Examples of strong acids include hydrochloric acid, which is produced in the stomach and helps break down food and kills bacteria, sulfuric acid and nitric acid.

Strong acids are completely ionised in water. When hydrochloric acid is placed in water, every hydrogen chloride molecule splits up to form hydrogen ions and chloride ions:

Hydrochloric acid → Hydrogen + Chlorine

$$HCl(g) + (aq) \rightarrow H^+(aq) + Cl^-(aq)$$

Build Your Understanding

Other acids are described as weak acids. Examples of weak acids include ethanoic acid, citric acid and carbonic acid.

Weak acids do not completely ionise in water. When ethanoic acid is placed in water, only a small fraction of the ethanoic acid molecules split up to form hydrogen ions and ethanoate ions:

$$CH_3COOH(l) \rightleftharpoons H^+(aq) + CH_3COO^-(aq)$$

Notice that this reaction is reversible. Ethanoic acid reacts more slowly with metals and carbonates than a comparative amount of a strong acid like hydrochloric acid would do. This is because ethanoic acid produces fewer H^+ ions and so there are fewer collisions between reactant particles and H^+ ions.

A sample of a weak acid, like ethanoic acid, has a lower electrical conductivity than a sample of a strong acid, like hydrochloric acid, because

hydrochloric acid is fully dissociated (that is, split up) in water and produces more H^+ ions.

However, both acids would produce the same volume of carbon dioxide if they were reacted with calcium carbonate or magnesium carbonate.

Electrolysis of both hydrochloric acid and ethanoic acid produces hydrogen gas at the negative electrode.

Weak acids, such as vinegar, are widely used as descalers since they remove limescale without damaging the surface of the object being cleaned.

Concentrated sulfuric acid is a dehydrating agent and can be used to remove water from sugar and from hydrated copper (II) sulfate.

✓ Maximise Your Marks

Acids are proton donors. Bases are proton acceptors.

Weak and Strong Alkalis

Traditional sources of alkalis included stale urine and burned wood. With industrialisation, the demand for alkalis grew, so shortages of alkalis soon developed.

Alkalis were used to neutralise acid soils, to produce the chemicals needed to bind dyes to cloth, to convert fats and oils into soap and to manufacture glass.

Weak and Strong Alkalis (cont.)

Early methods of manufacturing alkalis from limestone and salt produced a lot of **pollution**, including the acid gas hydrogen chloride and waste heaps that slowly released the toxic and unpleasant smelling gas hydrogen sulfide. **Oxidation** of hydrogen chloride forms chlorine gas.

Alkaline solutions have a pH more than 7. Some alkalis are described as **strong alkalis**. Examples of strong alkalis include sodium hydroxide and potassium hydroxide.

Strong alkalis are completely ionised in water. When sodium hydroxide is placed in water, it splits up to form sodium ions and hydroxide ions:

$$NaOH(s) + (aq) \rightarrow Na^+(aq) + OH^-(aq)$$

Other alkalis are described as weak. Ammonia is an example of a **weak alkali**.

Weak alkalis do not completely ionise in water. Ammonia produces hydroxide, OH^-, ions when it reacts with water:

Ammonia + Water ⇌ Ammonium + Hydroxide ion   ion

$$NH_3 + H_2O \rightleftharpoons NH_4^+ + OH^-$$

Ammonium salts are useful **fertilisers**.

✓ Maximise Your Marks

Hydroxide ions have the formula OH^-. Remember to add the negative charge. Strong alkalis have a high pH and are fully ionised.

The pH Scale

The pH scale can be used to distinguish between weak and strong acids and alkalis. The pH scale measures the concentration of hydrogen ions. Neutral solutions have a pH of 7. Acidic solutions have a pH of less than 7.

The strongest acids have a pH of 1. Dilute solutions of weak acids have higher pH values than dilute solutions of strong acids. If water is added to an acid it becomes more dilute and less corrosive.

Alkaline solutions have a pH of more than 7. The strongest alkalis have a pH of 14. Many cleaning materials contain alkalis. If water is added to an alkali it becomes more dilute and less corrosive.

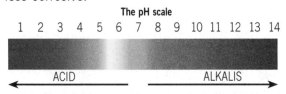

The pH scale

1 2 3 4 5 6 7 8 9 10 11 12 13 14

ACID ← → ALKALIS

Indicators

Indicators can be used to show the pH of a solution. Indicators work by changing colour. They can show when exactly the right amount of acid and alkali have been added together.

Build Your Understanding

Red litmus turns blue in alkaline conditions; blue litmus turns red in acidic conditions.

❓ Test Yourself

1. Which ions are found in acidic solutions?
2. Which ions are found in alkaline solutions?
3. Name three strong acids and explain why they are described as 'strong'.

⭐ Stretch Yourself

1. A 1 g sample of calcium carbonate was placed in 50 cm³ of 1.0 mol dm⁻³ ethanoic acid. An identical sample of calcium carbonate was then placed in 50 cm³ of 1.0 mol dm⁻³ hydrochloric acid.
   a) In what ways would the reactions be the same?
   b) In what ways would the reactions be different?

Making Salts

Neutralisation Reactions

The reaction between an acid and a base is called **neutralisation**:

- Acidic solutions contain hydrogen, H^+, ions.
- Alkaline solutions contain hydroxide, OH^-, ions.

The reaction between an acid and an alkali can be shown in the word equation:

Acid + Alkali → Salt + Water

The ionic equation for all neutralisation reactions is:

$H^+(aq) + OH^-(aq) \rightarrow H_2O(l)$

The type of salt that is produced during the reaction depends on the acid and the alkali used. Indigestion medicines contain chemicals that react with, and neutralise, excess stomach acid.

Naming Salts

Neutralising hydrochloric acid will produce **chloride salts**:

Hydrochloric	**+ Sodium**	**→ Sodium**	**+ Water**
Acid	Hydroxide	Chloride	

Neutralising nitric acid will produce **nitrate salts**:

Nitric	**+ Potassium**	**→ Potassium**	**+ Water**
Acid	Hydroxide	Nitrate	

Neutralising sulfuric acid will produce **sulfate salts**:

Sulfuric +	**Sodium**	**→ Sodium**	**+ Water**
Acid	Hydroxide	Sulfate	

Ammonia reacts with water to form a weak alkali. Ammonia solution can be neutralised with acids to form **ammonium salts**.

Build Your Understanding

Metal oxides are also bases. They can be reacted with acids to make salts and water:

Metal Oxide + Acid → Salt + Water

Examples

Copper	**+ Hydrochloric**	**→ Copper**	**+ Water**
Oxide	Acid	Chloride	

| **CuO** | **+** | **2HCl** | **→ $CuCl_2$** | **+ H_2O** |

Zinc	**+**	**Sulfuric**	**→**	**Zinc**	**+**	**Water**
Oxide		Acid		Sulfate		

| **ZnO** | **+** | **H_2SO_4** | **→** | **$ZnSO_4$** | **+** | **H_2O** |

Copper	**+**	**Nitric**	**→**	**Copper**	**+**	**Water**
Oxide		Acid		Nitrate		

| **CuO** | **+** | **$2HNO_3$** | **→** | **$Cu(NO_3)_2$** | **+** | **H_2O** |

pH Curves

The diagram on page 87 can be used to analyse what happens to the pH as acids and alkalis react together.

In this experiment, 25 cm³ of 0.1 mol dm⁻³ of the strong acid hydrochloric acid was placed in a flask and the pH was measured using a pH meter. A 0.1 mol dm⁻³ solution of the alkali sodium hydroxide was placed in a burette. A small amount of the alkali was added to the flask containing the acid and the new pH was recorded. This was repeated until all the alkali had been added.

pH Curves (cont.)

At first, as the alkali is added the pH changes very little as there is a large excess of the strong acid. Nearer to the **equivalence point** (when exactly the right amount of alkali has been added to react with all the acid) the pH increases more quickly.

Eventually there is a very sharp increase in pH as a tiny amount of alkali is added. The equivalence point is the middle of this vertical part of the graph. The graph shows the volume of alkali needed to react with all the acid. As more alkali is added the pH increases as the alkali is now in excess.

It is necessary to choose an indicator that changes colour suddenly over the vertical part of the graph; **phenolphthalein** changes colour between 8.5 and 10.0. At a low pH it is colourless; at higher pH it is pink. pH curves can also be drawn for acids being added to alkalis.

The effect on pH as acids and alkalis react together

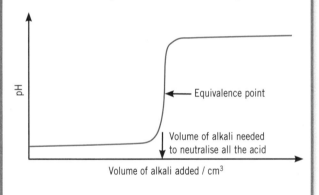

pH

Equivalence point

Volume of alkali needed to neutralise all the acid

Volume of alkali added / cm³

Making Salts from Metals

Some metals can be reacted with acids to form a salt and hydrogen.

Salts of very unreactive metals, such as copper, cannot be made in this way because these metals do not react with acids.

Salts of very reactive metals, such as sodium, cannot be made in this way because the reaction between the metal and acid is too vigorous to be carried out safely.

Precipitation Reactions

Some **insoluble** salts can be made from the reaction between two solutions. Barium sulfate is an insoluble salt. It can be made by the reaction between solutions of barium chloride and sodium sulfate:

Barium + Sodium → Barium + Sodium
Chloride   Sulfate   Sulfate   Chloride

Precipitation reactions can be used to remove unwanted ions from solutions. This technique is used to treat drinking water and effluent.

> ✓ **Maximise Your Marks**
>
> To get a top grade, make sure you can write the symbol equation for this reaction:
>
> $BaCl_2(aq) + Na_2SO_4(aq) \rightarrow BaSO_4(s) + 2NaCl(aq)$

❓ Test Yourself

1. What is formed when sulfuric acid reacts with sodium hydroxide?
2. What is the reaction between an acid and a base called?
3. Which ions are found in all acids?
4. What sort of salt does nitric acid produce?

⭐ Stretch Yourself

1. What is produced when acids react with metal oxides?
2. In a reaction an alkali is added to an acid. What is the meaning of the equivalence point in this reaction?

Limestone

Types of Rock

Rocks can be classified into three groups: **sedimentary**, **metamorphic** and **igneous**:

- **Limestone** and **chalk** are sedimentary rocks. These rocks sometimes contain **fossils** and are relatively soft and easy to erode. Sedimentary rocks are formed when layers of sediment are compacted over millions of years. The presence of shell fragments and ripple marks in fossils indicate that the rocks formed in a **marine environment**. The shapes of sand grains can also indicate the environment in which the rock formed.
- **Marble** is an example of a metamorphic rock. Marble is made when limestone or chalk are subjected to high pressures and temperatures.
- **Granite** is an example of an igneous rock. Igneous rocks are harder than metamorphic rocks which, in turn, are harder than sedimentary rocks. They are formed when **magma** (liquid rock below the Earth's surface) or **lava** (liquid rock on the Earth's surface) cool down and solidify. The faster the crystals in the rocks form the smaller the crystals will be.

Natural geological processes, such as **sedimentation** and **evaporation**, lead to the formation of valuable resources such as salt, limestone and coal.

Build Your Understanding

The materials used in everyday life, such as metals, ceramics and polymers, are chemicals or mixtures of chemicals. Some materials, such as cotton, paper, silk and wool, are made from living things.

Raw materials from the Earth's crust can be made into useful new synthetic materials. Chemical industries developed in the north-west of England because important resources including salt, limestone and coal could be found nearby.

When limestone (calcium carbonate) is heated, it breaks down to form quicklime (calcium oxide) and carbon dioxide:

Calcium Carbonate	$\rightarrow$	Calcium Oxide	+	Carbon Dioxide
$CaCO_3(s)$	$\rightarrow$	$CaO(s)$	+	$CO_2(g)$

This is an example of a thermal decomposition reaction (heat is used to break down the substance into simpler substances).

Quicklime can be reacted with water to form slaked lime (calcium hydroxide).

A solution of slaked lime is known as limewater:

Calcium Oxide + Water → Calcium Hydroxide

$$CaO(s) + H_2O(l) \rightarrow Ca(OH)_2(s)$$

Calcium carbonate, calcium oxide and calcium hydroxide are all bases and so can be used to neutralise acidic lakes and soils.

Limewater is used to test for the presence of the gas carbon dioxide. Carbon dioxide turns limewater cloudy:

Calcium Hydroxide	+	Carbon Dioxide	$\rightarrow$	Calcium Carbonate	+	Water

$$Ca(OH)_2(aq) + CO_2(g) \rightarrow CaCO_3(s) + H_2O(l)$$

When limestone is heated, its mass decreases because the gas carbon dioxide is produced and escapes.

Limestone quarry

The thermal decomposition of limestone is an example of a reaction that takes in heat. This is called an endothermic reaction. The formula $CaCO_3$ shows us the type and ratio of atoms present. Here, the calcium, carbon and oxygen atoms are present in the ratio 1 : 1 : 3. Calcium oxide is an example of a compound that is held together by ionic bonds. Ionic bonding involves the transfer of electrons. This forms ions with opposite charges, which then attract each other.

Interpreting Other Chemical Formulae

In chemistry it is very useful to interpret other formulae in terms of the number of each type of atom present.

Name	Formula	Type and Ratio of Atoms Present
Sodium hydroxide	NaOH	1 sodium : 1 oxygen : 1 hydrogen
Copper nitrate	$Cu(NO_3)_2$	1 copper : 2 nitrogen : 6 oxygen

Limestone as a Building Material

Limestone is used to make iron and steel and to build roads. It can also be used to make a range of building materials:

- **Cement** is produced by roasting powdered clay with powdered limestone in a rotating kiln. If water is added, and the mixture is allowed to set, it forms the hard, stone-like material cement. Clay can also be used to make bricks.
- When water is mixed with cement and sand, and then allowed to set, **mortar** is made.
- **Concrete** is made by mixing cement, sand and aggregate (rock chippings) with water. When water is added to cement it **hydrates** and binds together all the particles to form a material that is as hard as rock. Concrete is tough and cheap; it is widely used in building, for example, bridges.
- **Reinforced concrete** is a useful **composite material**. It is made by setting concrete around steel supports. This material combines the hardness of concrete with the flexibility and strength of steel.
- **Glass** can be made by heating up a mixture of limestone (calcium carbonate), sand (silicon dioxide) and soda (sodium carbonate) until the mixture melts.

Acids, Bases and Salts

? Test Yourself

1. What is the main chemical in limestone?
2. What type of rock is limestone?
3. What is magma?
4. Name three valuable resources made by natural geological processes.

★ Stretch Yourself

1. When calcium carbonate is heated fiercely a chemical reaction takes place.
   a) Write a balanced symbol equation for this reaction.
   b) Why does the sample of calcium carbonate have a lower mass after heating?

Metal Carbonate Reactions

Thermal Decomposition Reactions

When the metal carbonate calcium carbonate is heated fiercely it decomposes to form calcium oxide and carbon dioxide. The general equation for the reaction is:

Metal Carbonate → Metal Oxide + Carbon Dioxide

Other metal carbonates, including the carbonates of copper, iron, manganese, calcium, magnesium and zinc, decompose in a similar way when they are heated. When copper carbonate is heated it breaks down to give copper oxide and carbon dioxide. Copper carbonate is green while copper oxide is black. The colour change shows a new substance has been made.

Heating copper carbonate

Copper Carbonate → Copper Oxide + Carbon Dioxide

$$CuCO_3(s) \quad \rightarrow \quad CuO(s) \quad + \quad CO_2(g)$$
(green)   (black)

✓ Maximise Your Marks

Be able to give the symbol equation for this reaction:

$$CuCO_3 \rightarrow CuO + CO_2$$

Group 1 Carbonates and Naming Compounds

Not all metal carbonates of group 1 metals will decompose at the temperatures that can be reached using a Bunsen burner. When these metal carbonates are heated they do not break down.

If two elements join together in a chemical reaction, the name of the compound is given by the two elements that have joined together, for example:

Sodium + Chlorine → Sodium Chloride

Heating Baking Powder

When metal hydrogen carbonate compounds are heated, they undergo thermal decomposition reactions to form metal carbonates, carbon dioxide and water.

The main chemical compound in baking powder is sodium hydrogen carbonate, $NaHCO_3$. When heated it reacts to form sodium carbonate, carbon dioxide and water. It is the carbon dioxide produced by the reaction that makes cakes rise:

Sodium Hydrogen → Sodium + Carbon + Water
Carbonate   Carbonate   Dioxide

✓ Maximise Your Marks

Be able to recall the equation that sums up this reaction:

$$2NaHCO_3 \rightarrow Na_2CO_3 + CO_2 + H_2O$$

Metal Carbonates

Metal carbonates react with acids to form a salt, water and carbon dioxide:

Metal Carbonate + Acid → Salt + Water + Carbon Dioxide

A gas (carbon dioxide) is made so bubbles will be seen. The name of the salt produced depends on the acid and the metal carbonate used:

- Calcium carbonate makes calcium salts.
- Hydrochloric acid makes chloride salts.
- Sulfuric acid makes sulfate salts.
- Nitric acid makes nitrate salts.

Making Salts from Metal Carbonates

Acids can be neutralised by metal carbonates to form salts. Most metal carbonates are insoluble, so they are bases, but they are not alkalis. When acids are neutralised by metal carbonates, a salt, water and carbon dioxide are produced. This means that rocks, such as limestone, that contain metal carbonate compounds are damaged by acid rain.

The general equation for the reaction is:

Metal Carbonate + Acid → Salt + Water + Carbon Dioxide

The reactions between acids and metal carbonates are **exothermic**.

Build Your Understanding

Examples of reactions between acids and metal carbonates:

Zinc Carbonate + Sulfuric Acid → Zinc Sulfate + Water + Carbon Dioxide

$$ZnCO_3 + H_2SO_4 \rightarrow ZnSO_4 + H_2O + CO_2$$

Copper Carbonate + Hydrochloric Acid → Copper Chloride + Water + Carbon Dioxide

$$CuCO_3 + 2HCl \rightarrow CuCl_2 + H_2O + CO_2$$

Make sure you can apply this idea to other examples.

The diagram opposite shows how copper chloride salt is made.

The steps involved in the production of copper chloride are as follows:

- Copper carbonate is added to hydrochloric acid until all the acid is used up.
- Any unreacted copper carbonate is filtered off.
- The solution of copper chloride and water is poured into an evaporating basin.
- The basin is heated gently until the first crystals of copper chloride start to appear.
- The solution is then left in a warm place for a few days to allow the remaining copper chloride to crystallise.

Naming Other Salts

Ethanoic acid is neutralised to form ethanoate salts. Phosphoric acid is neutralised to form phosphate salts.

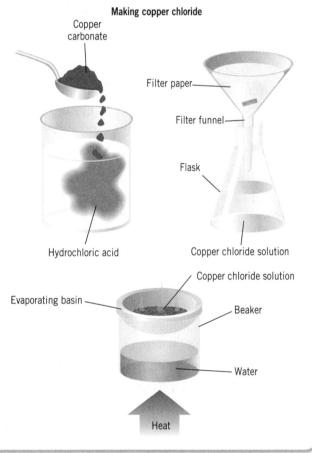

Making copper chloride

Copper carbonate

Filter paper

Filter funnel

Flask

Hydrochloric acid

Copper chloride solution

Copper chloride solution

Evaporating basin

Beaker

Water

Heat

❓ Test Yourself

1. Which gas is given off when metal carbonates react with acids?
2. Which sort of salts does hydrochloric acid make?
3. Which sort of salts does sulfuric acid make?
4. Which sort of salts does nitric acid make?

⭐ Stretch Yourself

1. A sample of copper carbonate was reacted with hydrochloric acid. Give the symbol equation for this reaction.

The Electrolysis of Sodium Chloride Solution

Sodium Chloride

Sodium chloride (**common salt**) is an important resource. It is an **ionic compound** formed from the **combination** of a group 1 metal (sodium) and a group 7 non-metal (chlorine). Sodium chloride is dissolved in large quantities in **seawater**.

Salt can be obtained by **mining** rocks or from allowing seawater to **evaporate**; the method used depends on how the salt is to be used and how pure it needs to be. **Quarrying** salt can have a dramatic impact on the environment. **Rock salt** (unpurified salt) is often used on icy roads. The salt lowers the freezing point of water from 0 °C to about −5 °C. Sprinkling rock salt on roads means that any water present will not freeze to form ice unless the temperature is very low.

Salt is also used in cooking, both to **flavour** food and as a **preservative**. However, eating too much salt can increase blood pressure and the chance of heart disease and strokes occurring. **Food packaging** may contain information on the sodium levels in a food.

Sodium ions may come from several sources, including sodium chloride salt. Government bodies, such as the Department for Health, produce guidelines to inform the public about the effects of different foods. A solution of sodium chloride in water is called brine.

To much salt can affect blood pressure

Electrolysis

The **electrolysis** of concentrated sodium chloride solution is an important industrial process and produces three useful products: chlorine, hydrogen and sodium hydroxide. The electrodes are made of inert materials so they do not react with the useful products made during the electrolysis reaction. The substance broken down is called the **electrolyte**.

Build Your Understanding

- During electrolysis, hydrogen ions, H^+, are attracted to the negative electrode where they pick up electrons to form hydrogen atoms, which pair up to make hydrogen, H_2, molecules:

  Hydrogen ions + Electrons → Hydrogen molecules

  $$2H^+ + 2e^- \rightarrow H_2$$

- Chloride ions, Cl^-, are attracted to the positive electrode where they release electrons to form chlorine atoms, which pair up to make chlorine molecules:

  Chloride ions − Electrons → Chlorine molecules

  $$2Cl^- - 2e^- \rightarrow Cl_2$$

- A solution of sodium hydroxide, $NaOH$, is also produced.

Each of these products can be used to make other useful materials. When there is a mixture of ions, such as in this case, the products that are formed depends on the **reactivity** of the elements involved. Electrolysis can also be used to **electroplate** objects. This can protect surfaces from **corrosion** and make them more attractive. Copper and silver plating are both produced by electrolysis.

Oxidation and Reduction

In the electrolysis of a concentrated sodium chloride solution:

- Hydrogen ions are **reduced** to hydrogen molecules; the hydrogen ions both gain an electron to form a hydrogen molecule.
- Chloride ions are **oxidised** to chlorine molecules; the two chloride ions both lose an electron to form a chlorine molecule.

Reduction and oxidation reactions must always occur together, so they are sometimes referred to as **redox** reactions.

Useful Products from the Electrolysis of Sodium Chloride

Chlorine is used:
- To make **bleach**.
- To sterilise water.
- To produce hydrochloric acid.
- In the production of PVC.

Hydrogen is used in the manufacture of margarine.

Sodium hydroxide is an alkali used in paper making and in the manufacture of many products including soaps and detergents, and rayon and acetate fibres.

Electrolysis of Molten Sodium Chloride

Solid sodium chloride does not conduct electricity because its ions cannot move. However, if sodium chloride is heated until it becomes **molten**, the sodium ions and chloride ions can move and electrolysis can occur.

Build Your Understanding

During the electrolysis of molten sodium chloride, the ions move towards the oppositely charged electrodes. Sodium, Na^+, ions (cations) are attracted to the negative electrode (cathode) where they pick up electrons to form sodium, Na, atoms:

Sodium ion + Electron → Sodium atom

$$Na^+ + e^- \rightarrow Na$$

Chloride, Cl^-, ions (anions) are attracted to the positive electrode (anode) where they release electrons to form chlorine atoms and then molecules:

Chloride ions − Electrons → Chlorine molecules

$$2Cl^- - 2e^- \rightarrow Cl_2$$

Make sure you can apply these ideas to other examples.

❓ Test Yourself

1. Which groups in the periodic table do sodium and chlorine belong to?
2. Where is sodium chloride found?
3. What is chlorine used for?
4. Why does solid sodium chloride not conduct electricity?
5. What is a solution of sodium chloride called?

⭐ Stretch Yourself

1. Give the symbol equations to show what happens at both the electrodes during the electrolysis of molten sodium chloride.
2. Name a health problem associated with eating too much salt.

Titrations

Titrations

In **titrations**, acid and alkali solutions are carefully added together. **Alkalis** are **soluble** bases. Titrations are really useful because they can be used to work out the **concentration** of one of the solutions used:

- Use a **pipette** to place a known amount of alkali into a flask. Pipettes measure out volumes very accurately. The bottom of the **meniscus** should lie exactly on the mark on the pipette.
- Place a couple of drops of **indicator** into the flask. Although many types of indicators are available, the two most widely used in titrations are methyl orange (which changes colour here from yellow to orange) and phenolphthalein (which changes from pink to colourless). Either of these indicators will work well when strong acids, such as hydrochloric acid, nitric acid or sulfuric acid, and strong bases, such as sodium hydroxide or potassium hydroxide, are used. Alternatively, a pH meter can be used to detect the end point of the titration. Universal indicator is not suitable because it is a mixture of several different indicators and it changes colour continuously; a single indicator that changes colour suddenly is essential.
- Place the acid in a **burette**.
- Add the acid into the flask gradually, a little at a time. Swirl the flask to ensure the solutions are fully mixed. When the indicator changes colour record the volume of acid that has been added. This is a rough titration to work out approximately how much acid to add.

- Repeat the titration. Using the results from the previous step, it should be possible to know roughly how much acid to use. Near the equivalence point, add the acid carefully, drop by drop, to find out exactly how much acid is required.
- Repeat the accurate titrations until you get two **concordant** results (results that are the same).

Titrations

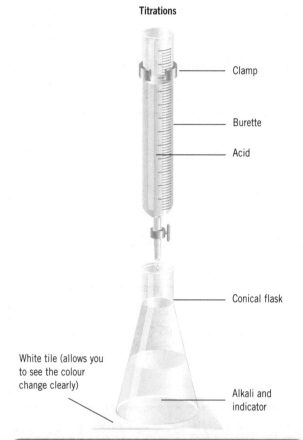

Clamp

Burette

Acid

Conical flask

White tile (allows you to see the colour change clearly)

Alkali and indicator

✔ Maximise Your Marks

Phenolphthalein is pink in alkalis and colourless in acids.

Build Your Understanding

Although an acid was added to an alkali in the example, it is possible to do the titration the other way round, placing the acid in the flask and the alkali in the burette. A sample of the salt can be obtained by repeating the titration with exactly the same volumes of the reactants but without the indicator. The reaction produces a salt and water. The water can be evaporated to leave crystals of the salt. Fertilisers can be made by reacting an acid with an alkali in this way.

Build Your Understanding

Example 1: 25 cm³ of a 0.1 mol dm⁻³ solution of sodium hydroxide was placed in a flask. 22.5 cm³ of hydrochloric acid was required for neutralisation. What is the concentration of the hydrochloric acid?

First, write down the balanced equation for the reaction and the values already known:

NaOH + HCl → NaCl + H₂O

25 cm³   22.5 cm³
0.1 mol dm⁻³   ?

Then work out the number of moles (amount) of sodium hydroxide:

Number of moles = $\frac{\text{volume}}{1000}$ × concentration

$\frac{25}{1000}$ × 0.1 = 0.0025 moles

From the balanced equation, 1 mole of sodium hydroxide reacts with 1 mole of hydrochloric acid, so 0.0025 moles of sodium hydroxide reacts with 0.0025 moles of hydrochloric acid.

Finally, work out the concentration of the acid:

Concentration = number of moles × $\frac{1000}{\text{volume}}$

= 0.0025 × $\frac{1000}{22.5}$

= 0.11 mol dm⁻³

The values from a number of titration reactions can be used to assess the degree of uncertainty in the values calculated.

Example 2: 25 cm³ of sodium hydroxide solution containing 0.1 g of sodium hydroxide pellets was reacted with 0.05 mol dm⁻³ of sulfuric acid. What volume of sulfuric acid would be required for neutralisation?

2NaOH + H₂SO₄ → Na₂SO₄ + 2H₂O

First, work out the number of moles of sodium hydroxide:

Number of moles = $\frac{\text{mass of sample}}{\text{formula mass of NaOH}}$

= $\frac{0.1}{40}$

= 0.0025 moles

From the balanced symbol equation, 2 moles of sodium hydroxide react with 1 mole of sulfuric acid, so 0.0025 moles of sodium hydroxide reacts with 0.00125 moles of sulfuric acid.

Finally, work out the volume of sulfuric acid required:

Volume = number of moles × $\frac{1000}{\text{concentration}}$

= 0.00125 × $\frac{1000}{0.05}$

= 25 cm³

✓ Maximise Your Marks

A strong alkali and a weak alkali of the same concentration require the same amount of acid for neutralisation.

✓ Maximise Your Marks

Try to set out your calculations in a logical order so the examiner can award marks for your working.

? Test Yourself

1. What is the name given to soluble bases?

2. Why is the flask swirled during a titration?

3. Phenolphthalein indicator is added to a solution of hydrochloric acid. What colour is the indicator in this solution?

4. Why is universal indicator unsuitable for finding the end point in a titration?

★ Stretch Yourself

1. 20.0 cm³ of a 0.10 mol dm⁻³ solution of sodium hydroxide was placed in a flask. 28.5 cm³ of hydrochloric acid was required for neutralisation. What is the concentration of the hydrochloric acid?

Acids, Bases and Salts

Water and Solubility

Water

The Earth's hydrosphere consists of its oceans, seas, lakes and rivers. It consists of water, dissolved salts and gases. Only appropriate sources away from polluted areas are chosen for drinking water. In the UK, water sources are found in lakes, rivers, aquifers (rock formations that contain water) and reservoirs.

Purifying Water

To purify water, sedimentation is used first to remove impurities. Next, the water is filtered to make it even cleaner. Finally, chlorine is added to the water to kill most of the microorganisms in the water. These tiny organisms could multiply quickly and cause disease, so their numbers must be brought down to acceptable levels.

Some scientists are concerned that even low levels of chlorine in water could cause health problems when the chlorine reacts with organic substances in the water. Some substances may still remain in the water even after purification. These can sometimes be poisonous so scientists regularly check water quality.

In some areas fluoride is added to drinking water. This helps to protect children's teeth from decay. However, some people are concerned about adding chemicals to water and about people's right to choose what is best for themselves and their families.

Water Pollution

Nitrate fertilisers can cause problems if they are washed into lakes or streams. Water can also be polluted by nitrate fertilisers, lead compounds from lead pipes and pesticides that have been sprayed near to water resources such as reservoirs.

Uses of Water

Water has many uses including as a coolant, a raw material and as a solvent.

Water is a particularly good solvent and dissolves most ionic compounds. The solubility of a substance (called the solute) in water can be worked out by measuring the number of grams of the substance that will dissolve in 100 g of water. The more grams of the substance that dissolve the more soluble it is. Generally, the higher the temperature of the water the more soluble a substance will become.

Seawater

Seawater contains large amounts of dissolved salts. As there is no overall charge for the salts in seawater, the charge on the ions can be used to work out the formula of the salt.

Metal Ions Present in Seawater	Non-metal Ions Present in Seawater
Sodium, Na^+	Bromide, Br^-
Potassium, K^+	Chloride, Cl^-
Magnesium, Mg^{2+}	Sulfate, SO_4^{2-}

The compound sodium sulfate contains sodium, Na^+, and sulfate, SO_4^{2-}, ions. For every two sodium ions one sulfate ion is required. The overall formula for the compound is Na_2SO_4.

Solubility

- All sodium, potassium and ammonium salts are soluble.
- All nitrates are soluble.
- Most chlorides are soluble (apart from silver and lead). Most sulfates are soluble (apart from lead, barium and calcium).
- Sodium, potassium and ammonium hydroxides and carbonates are soluble.

Solubility curves are used to show how many grams of a particular substance will dissolve in a given solvent. Notice how the solubility of both salts increases as the temperature increases.

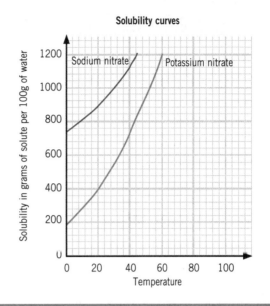

Solubility curves

Washing Powders

Washing powders are used to get clothes clean. They consist of many ingredients including:
- A detergent, to get rid of dirt.
- Water softeners, to remove the hardness from hard water so the detergent can work properly.
- Bleaches, to remove stains.
- Enzymes, to help remove stains at low temperatures.
- Optical brighteners, to make the clothes look very clean.

Clothes labels show symbols that indicate how the item should be washed. Washing clothes at lower temperatures than the one shown on the label will save money, but might not get the clothing as clean. More advanced washing powders work to clean clothes even at low temperatures. This helps save energy and money.

Build Your Understanding

Washing powders and washing-up liquids contain chemicals called detergents. Detergent molecules have two parts. Many detergent molecules form a sphere around the grease or fat, which can then be washed off.

Hydrophilic polar head group which forms bonds with water molecules

Hydrophobic tail which avoids water molecules but forms bonds with fat or grease

Dry cleaning
Some fabrics can be damaged by washing them in water. These fabrics should be dry-cleaned. Water is not involved in dry cleaning; other solvents are used to get the clothes clean. Some clothes can be damaged by stains that do not dissolve in water. However, these clothes can also be cleaned by dry-cleaning.

❓ Test Yourself

1. What are aquifers?
2. What is the role of a water softener in a washing powder?
3. Why are enzymes added to washing powders?

⭐ Stretch Yourself

1. Suggest two reasons why some clothes can only be successfully cleaned using the dry-cleaning method.

Hard and Soft Water

The Importance of Water

Water is an extremely important resource. The human body consists of about 70 per cent water and loses around two and a half litres of water every day. Most of this planet is covered in water, but the vast majority of this water is unsuitable to drink because it is in oceans and seas.

Hard and Soft Water

In some parts of the UK tap water is soft. Soft water does not contain dissolved calcium or magnesium salts. Distilled water, including fresh rainwater, can be described as being soft. Some tap water contains dissolved magnesium and calcium salts. This water can be described as being hard. As rainwater falls, carbon dioxide in the atmosphere dissolves in the water and then this solution reacts with calcium carbonate to form calcium hydrogen carbonate:

Carbon + Water + Calcium → Calcium
Dioxide   Carbonate   Hydrogen
                             Carbonate

$$CO_2(g) + H_2O(l) + CaCO_3(s) \rightarrow Ca(HCO_3)_2(aq)$$

Once the carbonic acid solution reaches the ground it moves through the soils and rocks. If these soils and rocks contain calcium or magnesium compounds, the carbonic acid solution will dissolve the compounds and so form hard water.

Types of Hard Water

Hard water can be classified as being permanently hard or temporary hard.

Permanently hard water contains calcium or magnesium chloride and sulfates. The calcium or magnesium ions remain in the solution even when the water is heated.

To soften hard water the calcium or magnesium ions need to be removed. One way to do this is to add sodium carbonate. Many washing powders contain this compound, also called washing soda. The sodium carbonate reacts with the calcium or magnesium ions to form precipitates of calcium or magnesium carbonate. As the calcium or magnesium ions have been removed and the water has been softened, the detergent in the washing powder will now work better:

Sodium   + Calcium → Sodium   + Calcium
Carbonate   Chloride   Chloride   Carbonate

$$Na_2CO_3(aq) + CaCl_2(aq) \rightarrow 2NaCl(aq) + CaCO_3(s)$$

Build Your Understanding

Soft water reacts readily with soap and water to form a lather. When hard water is reacted with soap it forms scum. The calcium (or magnesium) ions react with the stearate, $C_{17}H_{35}COO^-$, ions in the soap to form calcium (or magnesium) stearate or scum. If more soap is added, all the calcium and magnesium ions will eventually be removed and will form lather.

To find out how hard a sample of water is, add soap solution to a known volume of the sample and shake. The harder the water the more soap solution is required before a good lather is formed.

Soapless detergents, such as washing-up liquids, are particularly useful because they do not form scum even with hard water.

Build Your Understanding

Temporary hard water contains calcium or magnesium hydrogen carbonate. On heating, the hydrogen carbonate ions decompose to form carbonate ions. These ions react with the calcium or magnesium ions present to form precipitates, leaving the water softer:

$$Ca(HCO_3)_2(aq) \rightarrow CaCO_3(s) + H_2O(l) + CO_2(g)$$

Advantages and Disadvantages of Different Types of Water

Hard water has some health benefits:
- It helps to develop and maintain strong bones and teeth.
- It helps to protect against heart disease.

Soft water also has some advantages:
- Using soft water for cleaning reduces costs because less soap is needed.

Temporary hard water can cause limescale, which builds up on appliances such as kettles and in central heating systems. Over time, the amount of limescale increases and the efficiency of the appliance is reduced, so the appliance will cost more to run. Limescale can be removed with weak acids such as vinegar. These weak acids react with limescale to give a soluble calcium salt, water and carbon dioxide, but do not react with metals.

Weak acids are better than strong acids at cleaning limescale because they react with the limescale but not with the appliance itself, for example in a kettle.

Ion Exchange Columns

Ion exchange columns can also be used to soften hard water. The column contains beads coated in sodium or hydrogen ions. Hard water is passed through the column.

The calcium or magnesium ions in the hard water are exchanged for the sodium or hydrogen ions.

Some people prefer the taste of soft water or simply want to remove some of the substances dissolved in their tap water to improve its quality.

At home, water filters containing carbon, silver or ion exchange resins can all be used to remove some of the dissolved substances present in tap water.

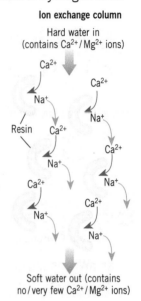

Ion exchange column

Hard water in (contains Ca^{2+} / Mg^{2+} ions)

Resin

Soft water out (contains no / very few Ca^{2+} / Mg^{2+} ions)

Desalination of Seawater

In desalination some of the salt and other minerals present in seawater are removed to produce fresh water. Pure water can be produced from seawater by distillation, but the process requires a great deal of energy to heat the impure water up to its boiling point. This makes the process too expensive to be worthwhile unless the fuel to be used is very cheap or the pure water is in particular demand.

Make sure you can draw the apparatus used to distill seawater.

❓ Test Yourself

1. What percentage of a human body is water?
2. Which types of salts make water hard?
3. What is the name of the chemical in washing soda?
4. Which types of ions are involved in an ion exchange column?

⭐ Stretch Yourself

1. Explain how you could identify whether a sample of water was soft, permanently hard or temporary hard.

Practice Questions

 Complete these exam-style questions to test your understanding. Check your answers on page 124. You may wish to answer these questions on a separate piece of paper.

1 5.0 g of calcium carbonate reacts with excess hydrochloric acid. The acid has a concentration of 1.0 mol dm^{-3}. The equation for the reaction is shown below.

$CaCO_3 + 2HCl \rightarrow CaCl_2 + H_2O + CO_2$

   a) How many moles of calcium carbonate were used in this reaction?   (1)

   b) Calculate the volume of hydrochloric acid required to react with all the calcium carbonate used in this reaction. Show your working.   (2)

2 In hard water areas, kettles and washing machines can be damaged by the build-up of limescale. Descaling products can be used to remove limescale from kettles. Many descalers contain weak acids.

   a) Why is limescale undesirable?   (1)

   b) What is a weak acid?   (2)

   c) Name a weak acid that could be used to remove limescale in the home.   (1)

   d) Why is a weak acid better than a strong acid for removing limescale from a kettle?   (1)

3 Water can be described as being hard or soft. Explain the difference between hard and soft water. Describe how hard water can be formed. What are the advantages of living in a hard water area and of living in a soft water area?

You should make sure your answers are written using good spelling, punctuation and grammar.   (6)

4 a) Complete the equation to show the products of the reaction between zinc carbonate and hydrochloric acid.

   Zinc Carbonate + Hydrochloric Acid → _____ + Water + Carbon Dioxide   (1)

b) What would you see during this reaction? (1)

c) How would you know when all the acid had been used up? (1)

d) How would you remove unreacted zinc carbonate? (1)

e) How would you get crystals of the salt? (1)

5 The electrolysis of concentrated sodium chloride solution is an important industrial process.

a) Name the gas produced at the positive electrode in this process. (1)

b) Give one use of this gas. (1)

c) Name the gas produced at the negative electrode in this process. (1)

d) Give one use of this gas. (1)

e) Name the other chemical made in this reaction. (1)

f) Give one use of this chemical. (1)

6 When sodium hydrogencarbonate is heated, it reacts to form sodium carbonate, carbon dioxide and water.

a) Balance this symbol equation to represent the reaction.

___$NaHCO_3 \rightarrow Na_2CO_3 + CO_2 + H_2O$ (1)

b) Which everyday substance, often found in a kitchen, contains sodium hydrogencarbonate? (1)

How well did you do?

| 0–10 | Try again | 11–17 | Getting there | 18–23 | Good work | 24–27 | Excellent! |

Relative Formula Mass and Percentage Composition

Calculation and Physical Chemistry

Why Relative Atomic Mass is Used

Relative atomic mass (RAM or A_r) is used to compare the masses of different atoms. The relative atomic mass of an element is the average mass of its **isotopes** compared with an atom of carbon-12.

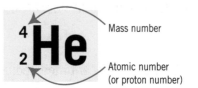

Mass number

Atomic number (or proton number)

Relative Formula Mass

The **relative formula mass (RFM or M_r)** of a substance is worked out by adding together the relative atomic masses of all the atoms in the ratio indicated by the formula.

Example 1: For nitrogen, N_2:

$$N_2$$
$$(2 \times 14) = 28$$

The relative formula mass of N_2 is 28. Nitrogen molecules contain a triple covalent bond, which is very strong. This makes nitrogen molecules very stable.

Example 2: For carbon dioxide, CO_2:

$$CO_2$$
$$12 + (2 \times 16) = 44$$

The relative formula mass of CO_2 is 44.

Example 3: For water, H_2O:

$$H_2O$$
$$(2 \times 1) + 16 = 18$$

The relative formula mass of H_2O is 18.

The **molar mass** of a substance is its relative formula mass in grams. The units for molar mass are **g/mol**.

Build Your Understanding

The relative formula mass of a substance in grams is known as 1 **mole** of the substance. This is also called the molar mass. 1 mole of CO_2 is 44 g and 1 mole of H_2O is 18 g. The number of moles of a substance present can be calculated using this formula:

$$\text{Number of moles} = \frac{\text{mass of sample}}{\text{relative formula mass of the substance}}$$

Example 1: How many moles are there in 9 g of water?

The relative formula mass of water is 18 so the molar mass is 18 g.

$$\text{Number of moles} = \frac{9}{18} = 0.5$$

There are 0.5 moles in 9 g of water.

Build Your Understanding (cont.)

Example 2: What is the mass of 0.5 moles of nitrogen, N_2?

The relative formula mass of nitrogen is 28.

$$\text{Mass of sample} = \text{number of moles} \times \text{relative formula mass of the substance}$$

$$= 0.5 \times 28 = 14 \text{ g}$$

The mass of 0.5 moles of nitrogen is 14 g.

Percentage Composition

Compounds consist of atoms of two or more different elements that have been chemically joined together. The percentage composition of an element in the compound can be calculated using this formula:

Percentage mass of an element in a compound $= \dfrac{\text{relative atomic mass} \times \text{no. of atoms}}{\text{relative formula mass}} \times 100\%$

Example 1: Ammonium nitrate is used as a fertiliser. Plants absorb fertiliser through their roots, so fertilisers must be soluble. Find the percentage composition of nitrogen in this compound:

- RAM of N = 14
- RAM of H = 1
- RAM of O = 16
- RAM of S = 32

The formula mass of ammonium nitrate, NH_4NO_3, is:

$$NH_4 \quad NO_3$$
$$14 + (4 \times 1) + 14 + (3 \times 16) = 80$$

Percentage of nitrogen $= \dfrac{14 \times 2}{80} \times 100\%$

$$= 35\%$$

The percentage of nitrogen in ammonium nitrate is 35 per cent.

Example 2: Ammonium sulfate is also used as a fertiliser. Find the percentage of nitrogen in this compound:

$$(NH_4)_2 \; SO_4$$
$$[14 + (4 \times 1)] \times 2 + 32 + (4 \times 16)$$

Percentage of nitrogen $= \dfrac{14 \times 2}{132} \times 100\%$

$$= 21.2\%$$

The percentage of nitrogen in ammonium sulfate is 21.2 per cent.

? Test Yourself

1 Why is relative formula mass used in science?

2 How is the relative formula mass of a substance calculated?

3 Find the relative formula mass of a nitrogen molecule, N_2.

4 Find the relative formula mass of carbon dioxide, CO_2.

★ Stretch Yourself

1 What is the relative formula mass of a substance in grams known as?

2 What is the mass of 1 mole of H_2O?

Calculating Masses

Calculating the Mass of Products

The masses of **products** and **reactants** can be worked out using the **balanced equation** for the reaction.

Example: What mass of water is produced when 8 g of hydrogen is burned?

Relative atomic masses:

H = 1, O = 16

First, write down what happens during the reaction as a word equation:

Hydrogen + Oxygen → Water

Then write it as a balanced symbol equation:

$2H_2 + O_2 → 2H_2O$

Next, calculate the **relative formula mass** of a hydrogen molecule and a water molecule.

The relative formula mass of hydrogen, H_2:

$$H_2$$
$$2 \times 1 = 2$$

The relative formula mass of water, H_2O:

$$H_2O$$
$$(2 \times 1) + 16 = 18$$

It is now possible to calculate the number of moles in 8 g of hydrogen.

$$\frac{8}{2} = 4 \text{ moles}$$

Next, examine the balanced symbol equation. Every 2 moles of hydrogen makes 2 moles of water. This means 4 moles of hydrogen will produce 4 moles of water.

Finally, work out the mass of 4 moles of water by rearranging the moles equation:

Mass of sample = number of moles × relative formula mass

$$= 4 \times 18$$
$$= 72$$

This shows that if 8 g of hydrogen is burned completely, 72 g of water vapour will be produced.

Build Your Understanding

The equation for a reaction can also be used to calculate how much of the reactants should be used to produce a given amount of the product.

Example: What mass of magnesium should be used to produce 60 g of magnesium oxide? Relative atomic masses:

Mg = 24, O = 16

First, write down what happens during the reaction as a word equation:

Magnesium + Oxygen → Magnesium Oxide

Then write it as a balanced symbol equation:

$2Mg + O_2 → 2MgO$

Next, calculate the relative formula mass of magnesium oxide.

The relative formula mass of magnesium oxide, MgO:

$$MgO$$
$$24 + 16 = 40$$

It is now possible to calculate the number of moles in 60 g of magnesium oxide.

$$\frac{60}{40} = 1.5 \text{ moles}$$

Build Your Understanding (cont.)

Next, examine the balanced symbol equation. To make 2 moles of magnesium oxide, 2 moles of magnesium are needed. So, to make 1.5 moles of magnesium oxide, 1.5 moles of magnesium are needed.

Finally, work out the mass of 1.5 moles of magnesium oxide by rearranging the moles equation.

Mass of sample = number of moles × relative formula mass

= 1.5 × 24

= 36

This shows that to make 60 g of magnesium oxide, 36 g of magnesium should be burned.

Percentage Yield

The amount of product made in a reaction is called the **yield**. Although atoms are never gained or lost during a chemical reaction, the yield of a reaction can be less than predicted:

- The reaction is **reversible** and does not go to completion.
- Some of the product is lost during **filtering**, **evaporation**, when transferring liquids or during **heating**.
- There may be **side-reactions** occurring that produce other products.

The amount of product actually made compared with the maximum calculated yield is called the **percentage yield**. A 100 per cent yield means no product has been lost; a 0 per cent yield means no product has been made:

$$\text{Percentage yield} = \frac{\text{mass of product}}{\text{maximum calculated yield}} \times 100\%$$

Scientists try to choose reactions with a high percentage yield or **high atom economy**. This contributes towards **sustainable development** by reducing waste. A 100 per cent atom economy means that all the reactant atoms have been made into the desired products.

Waste products are undesirable as they cannot be sold for profit, their disposal can be costly and cause environmental and social problems.

Ethanol

Ethanol can be made by two different methods with very different atom economies.

When ethanol is made by reacting ethene with steam, the process has an atom economy of 100 per cent:

$$C_2H_4 + H_2O \rightarrow C_2H_5OH$$

When ethanol is made by fermentation, the atom economy is less:

$$C_6H_{12}O_6 \rightarrow 2C_2H_5OH + 2CO_2$$

$$\text{Atom economy} = \frac{M_r \text{ of desired products}}{M_r \text{ of all products}} \times 100$$

$$= \frac{92}{180} \times 100 = 51.1\%$$

Calculation and Physical Chemistry

? Test Yourself

1. What mass of water vapour is produced when 4 g of hydrogen is burned?
2. What mass of water vapour is produced when 16 g of hydrogen is burned?

★ Stretch Yourself

1. Consider the equation below:
$$CaCO_3 \rightarrow CaO + CO_2$$
If 5.0 g of calcium carbonate is heated fiercely, what mass of calcium oxide is produced?

105

Calculations

Finding the Empirical Formula

The **empirical formula** of a compound is the simplest whole number ratio of the atoms it contains.

Example: Find the empirical formula of magnesium oxide formed when 12 g of magnesium reacts with 8 g of oxygen. Deal with the magnesium and oxygen separately.

The simplest ratio of magnesium atoms to oxygen atoms is 1 : 1 so the empirical formula is MgO.

	Mg	O
State the number of grams that combine	12	8
Change the grams to moles (divide by A_r)	$\dfrac{12}{24}$	$\dfrac{8}{16}$
This is the ratio in which the atoms combine	0.5	0.5
Get the ratio into its simplest form	1	1

Molar Mass

The **molar mass** is the mass of 1 mole of a substance in grams.

The number of moles in a sample of a

$$\text{substance} = \frac{\text{mass of sample}}{\text{molar mass}}$$

Example: How many moles are present in 60 g of calcium carbonate, $CaCO_3$?

The molar mass of calcium carbonate:

$$\underset{40}{\text{Ca}} \quad \underset{12}{\text{C}} \quad \underset{(16 \times 3) = 100}{\text{O}_3}$$

The number of moles in a sample of a

$$\text{substance} = \frac{\text{mass of sample}}{\text{molar mass}}$$

$$= \frac{60 \text{ g}}{100 \text{ g}}$$

$$= 0.6 \text{ moles}$$

Avogadro's Number

One mole of any substance contains 6×10^{23} particles. This is known as **Avogadro's number**. It can be used to work out how many moles of a substance are present.

$$\frac{\text{Number}}{\text{of moles}} = \frac{\text{number of particles in the sample}}{\text{number of particles in one mole}}$$

Example: A sample contains 4.5×10^{23} particles. How many moles are there in this sample?

$$\frac{\text{Number}}{\text{of moles}} = \frac{\text{number of particles in the sample}}{\text{number of particles in one mole}}$$

$$= \frac{4.5 \times 10^{23}}{6 \times 10^{23}}$$

$$= 0.75 \text{ moles}$$

Build Your Understanding

One mole of any gas occupies a volume of 24 dm³ at room temperature and pressure. This is known as the molar volume of a gas.

Volume in dm³ = number of moles × 24

Example 1: What is the volume of 1.5 moles of carbon dioxide, CO_2?

Volume in dm³ = number of moles × 24
= 1.5 × 24
= 36 dm³

Example 2: A sample of gas at room temperature and pressure occupies 40 dm³. How many moles of gas are present?

Number of moles = $\dfrac{\text{volume in dm}^3}{24}$

= $\dfrac{40}{24}$

= 1.67 moles

Concentration

The **concentration** of a solution is a measure of how much **solute** is dissolved in 1 dm³ of solution.
It is sometimes used to describe how many moles are dissolved in 1 dm³ of solution and has the units **mol dm⁻³**.

Concentration = $\dfrac{\text{moles}}{\text{volume in dm}^3}$

Example: If 2 moles of sodium hydroxide pellets are added to distilled water and the total volume of solution is 2 dm³, what is the concentration of the solution?

Concentration = $\dfrac{\text{moles}}{\text{volume in dm}^3}$

= $\dfrac{2 \text{ moles}}{2 \text{ dm}^3}$

= 1 mol dm⁻³

✓ Maximise Your Marks

Concentration is sometimes used to describe how many grams of a solute are dissolved in 1 dm³ of solution and has the units **g/dm³** or **g dm⁻³**.

It is often very useful to **dilute** concentrated solutions such as medicines, baby milk and orange cordial drinks. To dilute a 1.0 mol dm⁻³ solution to a 0.1 mol dm⁻³ solution, take 10 cm³ and make it up to 100 cm³ with water in a volumetric flask. There are 1000 cm³ in 1 dm³.

Concentrated solutions need to be diluted

? Test Yourself

1. A sample of magnesium contains 3×10^{23} atoms. How many moles of magnesium are in this sample?

2. A sample of iron contains 6×10^{23} atoms. How many moles of iron are in this sample?

3. How many atoms are in 0.1 moles of argon?

4. How many atoms are there in 0.5 moles of sodium?

★ Stretch Yourself

1. What volume does 12 moles of carbon dioxide occupy?

2. A 750 cm³ sample of sodium hydroxide solution contains 3 g of solid sodium hydroxide. What is the concentration of this solution in g/dm³?

Rates of Reaction

Slow and Fast Reactions

The **rate of reaction** is equal to either:
- The amount of reactant used up divided by the time taken.
- The amount of product made divided by the time taken.

Rusting is an example of a reaction that happens very slowly, while **combustion** reactions and explosions happen very quickly. **Explosions** produce a large volume of **gaseous products**. Factories that produce fine powders, such as custard powder, have to be very careful to prevent explosions from occurring.

Measuring Rates of Reaction

The method chosen to follow the rate for a particular reaction depends on the reactants and products involved. When sodium thiosulfate reacts with hydrochloric acid, one of the products is a precipitate of sulfur. The rate of reaction can be followed using a light sensor and a data logger to measure how quickly sulfur is being made.

A chemical reaction can only occur if the reacting particles collide with enough energy to react. This is called the **activation energy**. If the particles collide but do not have the minimum energy to react, the particles just bounce apart without reacting.

Measuring the rate of reaction

Gas syringe

Dilute hydrochloric acid

Magnesium

The rate of a chemical reaction can be measured by:
- How fast the products are being made.
- How fast the reactants are being used up.

The graph shows the amount of product made in two experiments. The lines are steepest at the start of the reaction in both experiments. The lines start to level out as the reactants get used up.

When the line becomes horizontal the reaction has finished. The graph shows that experiment A has a faster rate of reaction than experiment B. However, both experiments produce the same amount of product.

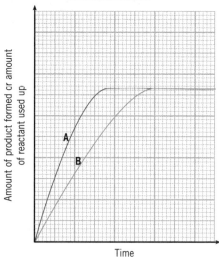

A graph to show how quickly a product is made in a chemical reaction.

Amount of product formed or amount of reactant used up

A

B

Time

When analysing graphs, the reaction is over when the graph levels out. Reactions stop when one of the reactants is all used up; this reactant is called the limiting reactant. The other reactants may not be completely used up and are said to be in excess. The amount of product made depends on the amount of reactant used up; the amount of product made is directly proportional to the amount of reactant used. The rate of reaction is measured using units of g/s or g/min or cm^3/s or cm^3/min.

Temperature

If the temperature is increased, the reactant particles move more quickly. Increasing the temperature increases the rate of reaction because:
- The particles collide more often.
- The particles have more energy when they collide.
- When they collide, the collision is more likely to lead to a reaction taking place between them.

Increasing the Surface Area

For a reaction to occur, the particles have to **collide**. The greater the surface area the more chance of the reactant particles colliding and the faster the rate of reaction.

With a small surface area (large pieces) the rate of reaction is slow. The particles collide less often. With a large surface area (small pieces) the rate of reaction is higher. The particles collide more often.

Large Particles	Small Particles
• Small surface area • Fewer collisions • Reaction rate is slow	• Large surface area • More collisions • Reaction rate is faster

✓ Maximise Your Marks

Remember, small pieces have a large surface area. The dust caused by fine powders, such as custard powder or flour, can burn explosively because of the large surface area of its particles.

Catalysts

A **catalyst** increases the rate of reaction, but is not itself used up during the reaction. Only a small amount of catalyst is needed to catalyse a large amount of reactants. Catalysts are specific to certain reactions. Reactions stop when one of the reactants is all used up. Catalysts offer an alternative reaction pathway with a lower activation energy.

Concentration and Pressure

For a reaction to take place, the reactant particles have to collide. If the concentration is increased there are more reactant particles in the solution. Increasing the concentration increases the rate of reaction because the particles collide more often.

For gases, increasing the pressure has the same effect as increasing the concentration of dissolved particles in solutions. At low pressure the rate of reaction slows down because the particles collide less often. At higher pressure the rate of reaction speeds up because the particles collide more often.

The concentrations of solutions are given in units of moles per cubic decimetre, $mol\ dm^{-3}$. Equal volumes of solutions of the same molar concentration contain the same number of moles of solute.

Increasing the pressure of gases increases the rate of reaction

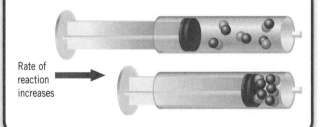

Rate of reaction increases

Calculation and Physical Chemistry

❓ Test Yourself

1. What happens to the rate of reaction if the concentration of reactants is increased?
2. What happens to the rate of reaction if the pressure of gaseous reactants is increased?
3. How does a catalyst affect the rate of reaction?

⭐ Stretch Yourself

1. Explain two ways in which increasing the temperature increases the rate of a chemical reaction.

Reversible Reactions

Simple Reversible Reactions

Not all reactions go to completion. Many chemical reactions are **reversible**; they can proceed both forwards and backwards.
If A and B are reactants and C and D are products, a reversible reaction can be summed up as:

A + B ⇌ C + D

The two reactants, A and B, can react to make the products C and D; at the same time, C and D can react together to produce A and B.

Dynamic Equilibrium

If a reversible reaction takes place inside a closed system (where nothing can enter or leave), an **equilibrium** will eventually be reached.

It will be a **dynamic equilibrium**: both the forwards and the backwards reactions are taking place at exactly the same rate.

The conditions will affect the position of equilibrium, that is, how much reactant and product are present at equilibrium.

If the forwards reaction is exothermic then increasing the temperature will decrease the amount of product made. If the forwards reaction is endothermic then increasing the temperature will increase the amount of product made.

Build Your Understanding

In a dynamic equilibrium, both the forwards and the backwards reactions are still happening. As they happen at the same rate there is no overall change in the concentrations of the reactants or the products.

Build Your Understanding (cont.)

If the forwards reaction is exothermic (gives out energy), then the backwards reaction is endothermic (takes in energy). The amount of energy given out by the forwards reaction must be the same as the amount of energy taken in by the backwards reaction.

Example: Hydrated copper (II) sulfate reactions

First, the hydrated (with water) copper sulfate is heated to make anhydrous (without water) copper sulfate. Then water is added to the anhydrous copper sulfate to produce hydrated copper sulfate.

Copper (II) sulfate reactions

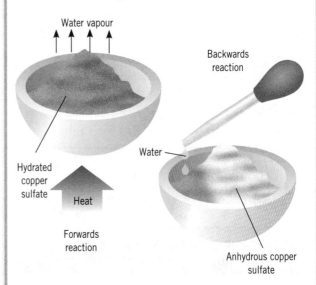

Water vapour

Backwards reaction

Hydrated copper sulfate

Heat

Water

Forwards reaction

Anhydrous copper sulfate

In the forwards reaction, the hydrated copper sulfate takes in energy as it is heated. This is an endothermic reaction:

Hydrated Copper → Anhydrous Copper + Water
Sulfate (blue)   Sulfate (white)

In the backwards reaction, energy is given out when water is added to the anhydrous copper sulfate. This is an exothermic reaction:

Anhydrous Copper + Water → Hydrated Copper
Sulfate (white)   Sulfate (blue)

Reactions Involving Gases

If a reaction involves gases then the pressure may affect the **yield** of the reaction.

First, count the number of gas molecules on the left-hand side and the right-hand side of the equation:

Reactants → Products

Fewer gas molecules → More gas molecules

Increasing the pressure decreases the yield of the product:

Reactants → Products

More gas molecules → Fewer gas molecules

Increasing the pressure increases the yield of the product.

Contact Process

The Contact process is used in the manufacture of sulfuric acid, H_2SO_4.

First, sulfur is burned in air to produce sulfur dioxide:

$S(s) + O_2(g) \rightarrow SO_2(g)$

Then the sulfur dioxide is reacted with more oxygen to form sulfur trioxide.

$2SO_2(g) + O_2(g) \rightleftharpoons 2SO_3(g)$

The forward reaction is exothermic.

Reaction Conditions
The raw materials are sulfur, water and oxygen. The reaction is carried out under the following conditions:
- Vanadium (V) oxide.
- Pressure of 1 atmosphere.
- Temperature of 450 °C.

Build Your Understanding

The vanadium (V) oxide catalyst is used because it increases the rate of reaction, which helps to reduce production costs.

A moderate temperature of 450 °C is chosen. This gives both a reasonable rate of reaction and a reasonable yield of the product.

The forwards reaction is exothermic so a higher temperature would give a faster rate of reaction but a lower yield of sulfur trioxide.

A lower temperature would give a higher yield of sulfur trioxide but a lower rate of reaction.

A higher pressure would increase the yield of the reaction because it would favour the forwards reaction, which decreases the number of gaseous molecules. However, it is expensive to maintain high pressures and, as the yield is already around 95 per cent, it is not necessary.

Finally, the sulfur trioxide is reacted with sulfuric acid to produce oleum, $H_2S_2O_7$:

$SO_3 + H_2SO_4 \rightarrow H_2S_2O_7$

The oleum is then reacted with water to form more sulfuric acid:

$H_2S_2O_7 + H_2O \rightarrow 2H_2SO_4$

? Test Yourself

1. What is special about a reversible reaction?
2. What is a closed system?
3. What is a dynamic equilibrium?

★ Stretch Yourself

1. The Contact process is used in the production of sulfuric acid.
   a) Name the catalyst used in the Contact process.
   b) Explain why a moderate temperature of 450 °C is used in this reaction.

The Haber Process

Making Ammonia

Ammonia is produced by the **Haber process** and is made from **nitrogen** and **hydrogen**.

Hydrogen is obtained from natural gas or from the **cracking** of oil fractions. Nitrogen is obtained from the **fractional distillation** of liquid air. The Haber process is an example of a **reversible reaction**:

$$N_2(g) + 3H_2(g) \rightleftharpoons 2NH_3(g)$$

Some of the nitrogen and the hydrogen react to form ammonia. At the same time, some of the ammonia breaks down into nitrogen and hydrogen.

On cooling, the ammonia **liquefies** and is removed from the reaction mixture.

Any unreacted nitrogen and hydrogen can be recycled to reduce costs. Ammonia is made on a very large scale.

Build Your Understanding

The cost of producing ammonia depends on:
- The cost of the raw materials.
- Energy costs.
- Equipment costs.
- Labour costs – the more automation the lower the wages bill will be.
- How quickly the ammonia is produced.

The industrial conditions are specially chosen. Typical conditions are:
- A high pressure (200 atmospheres).
- A moderate temperature (450 °C).
- An iron catalyst.

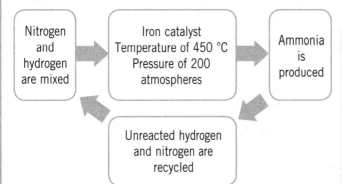

✓ Maximise Your Marks

To get a top grade, you need to explain the conditions used in the Haber process. It is not enough just to state that compromise conditions are used. You should explain why it is a compromise in terms of the rate of reaction and the yield of ammonia.

Choosing the Conditions for Producing Ammonia

A **high pressure** is used to increase the amount of ammonia produced. In the balanced symbol equation, there are four gas molecules on the left-hand side of the equation (one nitrogen molecule and three hydrogen molecules), but there are only two ammonia molecules on the right-hand side of the equation. Increasing the pressure encourages the forwards reaction, which increases the amount of ammonia produced because there are fewer gas

molecules on the right-hand side of the equation. Ideally, the highest possible pressures should be used. However, in practice, it is too expensive to build a plant that can withstand pressures greater than 200 atmospheres.

The reaction between nitrogen and hydrogen to produce ammonia is **exothermic**. A low temperature would increase the **yield** of ammonia produced at equilibrium.

Choosing the Conditions for Producing Ammonia (cont.)

It would, also make the rate of the reaction very slow. A higher temperature would give a much faster rate of reaction but the yield of ammonia at **equilibrium** would be much lower. In practice, a compromise temperature of 450 °C is used to give a reasonable yield of ammonia fairly quickly.

An iron catalyst is used to increase the rate of reaction and allows a lower temperature to be used. This helps to reduce the costs of making ammonia. The catalyst is not used up during the reaction so it can be used many times. Different catalysts work for different reactions but **transition metals** and their **compounds** are often good catalysts.

Build Your Understanding

Ammonia can be oxidised to produce nitric acid. Ammonia gas reacts with oxygen in the air using a platinum catalyst:

$$4NH_3 + 5O_2 \rightarrow 4NO + 6H_2O$$

The nitrogen oxide is cooled and then reacted with water and more oxygen to form nitric acid:

$$4NO + 3O_2 + 2H_2O \rightarrow 4HNO_3$$

Ammonium Nitrate
Nitric acid can be neutralised by ammonia to make ammonium nitrate:

Ammonia + Nitric Acid → Ammonium Nitrate

Make sure you can apply this idea to the production of ammonium phosphate, ammonium sulfate and potassium nitrate.

Making Fertilisers

Fertilisers replace the essential elements used by plants as they grow, so the manufacture of ammonia is important for the world's food production. Many fertilisers contain nitrogen, which is needed for plant growth. Commonly used artificial fertilisers include ammonium nitrate, ammonium phosphate and ammonium sulfate.

Plants absorb the chemicals in these fertilisers through their roots, so fertilisers must be soluble in water. Phosphorus and potassium are also needed for plants to grow well.

Fertilisers are sprayed on crops to promote growth

Fixing Nitrogen

Some living organisms can **fix** (that is, hold on to) nitrogen from the air at room temperature and pressure by using **enzymes**. Scientists have found that they can alter nitrogen **fixation levels** by using different catalysts. Scientists are now interested in developing new, more efficient catalysts that behave like the enzymes found in living organisms. Some ammonia is used to produce cleaning fluids.

❓ Test Yourself

1. What does the Haber process produce?
2. From where is the hydrogen obtained?
3. Why is the Haber process described as reversible?

⭐ Stretch Yourself

1. Explain how ammonia can be made into ammonium nitrate. Include symbol equations to sum up each step in the production of this fertiliser.

Exothermic and Endothermic Reactions

Calculation and Physical Chemistry

Energy Changes and Chemical Reactions

During chemical reactions atoms are rearranged as old **bonds** are broken and new bonds are made. **Energy** is required to break bonds and is released when new bonds are formed.

If, overall, energy is given to the surroundings, the reaction is described as **exothermic**. If, overall, energy is taken from the surroundings, the reaction is described as **endothermic**.

To work out whether a reaction is endothermic or exothermic, scientists measure the temperature of the chemicals before the reaction and then after the reaction. If the temperature has increased the reaction is exothermic; if the temperature has decreased the reaction is endothermic.

Exothermic and Endothermic Reactions

Burning methane is an example of an exothermic reaction. Rusting, explosions and neutralisation reactions are also exothermic. Self-heating cans and hand warmers make use of exothermic reactions.

The thermal decomposition of limestone is an example of an endothermic reaction. Photosynthesis and dissolving ammonium nitrate in water are also endothermic processes. Some sports injury packs make use of endothermic reactions.

Bond Energy Calculations

Each chemical bond has a specific **bond energy**. This is the amount of energy that must be taken in to break 1 mole of those bonds.

Bond	Bond energy (kJ mol^{-1})
C–H	413
O=O	496
C=O	743
O–H	463
C–O	358

Example: Burning the fuel methane, CH_4.

$$\begin{array}{ccc} \overset{\displaystyle H}{\underset{\displaystyle H}{H-C-H}} \; + \; \begin{array}{c} O=O \\ \\ O=O \end{array} & \rightarrow & O=C=O \; + \; \begin{array}{c} H\!\diagdown\!\!O\!\diagup\!H \\ H\!\diagup\!\!O\!\diagdown\!H \end{array} \end{array}$$

Energy taken in to break the bonds:
- 4 moles of C–H = 4 × 413 = 1652 kJ mol^{-1}
- 2 moles of O=O = 2 × 496 = 992 kJ mol^{-1}
- Total = 1652 + 992 = 2644 kJ mol^{-1}

Energy taken in has a positive sign (endothermic).

Energy given out when forming bonds:
- 2 moles of C=O = 2 × 743 = 1486 kJ mol^{-1}
- 4 moles of O–H = 4 × 463 = 1852 kJ mol^{-1}
- Total = 1486 + 1852 = 3338 kJ mol^{-1}

Energy given out has a negative sign (exothermic).

Difference in energy between the energy given out and the energy taken in:
= +2644 kJ mol^{-1} − 3338 kJ mol^{-1}
= −694 kJ mol^{-1}

This reaction gives out more energy than it takes in so it is exothermic. The burning of fuels is always exothermic.

Build Your Understanding

In an exothermic reaction, the products have less energy than the reactants. The difference in energy between the products and the reactants is the amount of energy given out by the reaction.

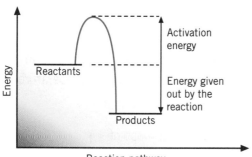

In an endothermic reaction, the products have more energy than the reactants. The difference in energy between the products and the reactants is the amount of energy taken in by the reaction.

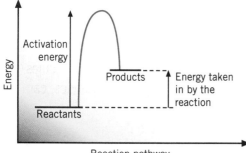

Catalysed Exothermic Reaction

More reactant particles will have the lower activation energy, so catalysed reactions happen faster. Catalysts increase the rate of reaction but are not used up during the reaction.

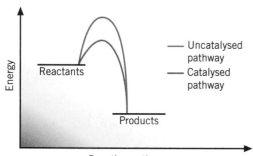

Catalysts provide an alternative reaction pathway that has lower activation energy.

Activation Energy

The **activation energy** is the minimum amount of energy needed to get a **reaction** started. This energy is needed to break the bonds in the reactants.

Catalysts provide an alternative reaction pathway that has lower activation energy. It is very important that chemists continue to develop better catalysts that lower activation energy.

By using a catalyst, it is possible to increase the rate of reaction. This means lower temperatures and pressures can be used, which has economic benefits, because it costs less, and environmental benefits, as high temperatures and pressures are often obtained by using electricity generated at power stations that burn fossil fuels. If less fossil fuel is being burned then less carbon dioxide is being released into the atmosphere. Using better catalysts also helps to preserve raw materials.

Power station

? Test Yourself

1 The temperature increases during a chemical reaction. What sort of reaction has taken place?

2 How do you know that burning coal is an exothermic reaction?

3 When hydrochloric acid neutralises sodium hydroxide the temperature increases. Is this an exothermic reaction or an endothermic reaction?

⭐ Stretch Yourself

1 Sketch an energy level diagram for the combustion of methane.

Explaining Energy Changes

Calorimetry – Burning Fuels

In **exothermic** reactions, more energy is given out when new bonds are formed than when the old bonds were broken. In **endothermic** reactions, more energy is taken in to break bonds than is released when new bonds are formed.

Scientists use **calorimetry** to compare the amount of energy released when fuels and foods are burned.

A measured amount of water is placed into a boiling tube and its temperature is measured using a thermometer. The fuel sample is then burned under the boiling tube containing the water, and the water is gradually warmed up. The temperature of the water is measured at the end of the experiment when the flame is extinguished.

The chemical energy that had been stored in the sample is released as thermal energy when the sample is burned. The greater the change in the temperature of the water the more energy was stored in the fuel.

Calorimetry

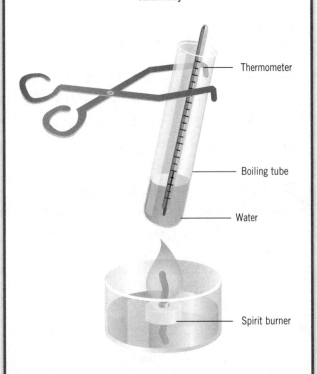

Thermometer

Boiling tube

Water

Spirit burner

Build Your Understanding

Specific heat capacity (SHC) is the amount of energy needed to increase the temperature of 1 g of the substance by 1 °C.

Energy = mass × specific heat × change in change   capacity   temperature

The SHC of water is 4.18 J/g/°C.

1 cm³ of water has a mass of 1 g.

Example: A 2 g sample of food made the temperature of 5 cm³ of water increase by 6 °C.

Energy = mass × specific heat × change in
change   capacity   temperature

$$= 5\ g \times 4.18\ J/g/°C \times 6\ °C$$

$$= 125.4\ J$$

In these equations the mass is the mass of the water, not the mass of the fuel or the chemicals used.

Although energy is usually measured in joules, other units, such as kilojoules or calories, can be used:

- 1000 joules = 1 kilojoule
- 1 calorie = 4.2 joules

To be able to compare the energy content of different samples, energy values are normally given for the same amount of substance: joules per gram, kilojoules per gram or calories per gram.

The energy values for the example are given below using different units:

$$\text{Joules per gram} = \frac{125.4\ J}{2\ g} = 62.7\ J/g$$

$$\text{Kilojoules per gram} = \frac{62.7\ J}{1000} = 0.0627\ kJ$$

$$\text{Calories per gram} = \frac{62.7\ J}{4.2\ J} = 14.9\ \text{calories}$$

Build Your Understanding (cont.)

To compare the energy transferred when liquid fuels are burned, place the fuel in a spirit burner. Measure the mass of the spirit burner and the fuel. Use the spirit burner to heat the water, then measure the new mass of the remaining fuel and the spirit burner. The amount of fuel used is the difference between the mass at the start and at the end.

💡 Boost Your Memory

Remember, combustion reactions are exothermic. Try making a list of all the other exothermic reactions you have learned about.

Calorimetry in Solutions

Calorimetry experiments can be used to work out the energy released by chemical reactions in solution. In the example below, a more reactive metal, iron, displaces a less reactive metal, copper, from a solution of copper sulfate.

Temperature at the start of the reaction	15 °C
Temperature at the end of the reaction	33 °C
Change in temperature	18 °C

The specific heat capacity of copper sulfate solution = 4.18 J/g/°C.

Measuring the temperature change of a displacement reaction

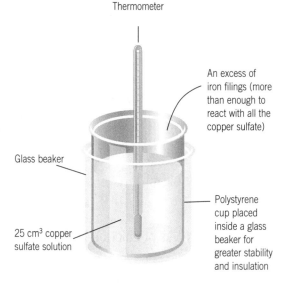

Thermometer

An excess of iron filings (more than enough to react with all the copper sulfate)

Glass beaker

Polystyrene cup placed inside a glass beaker for greater stability and insulation

25 cm³ copper sulfate solution

Build Your Understanding

The reaction below is carried out inside an insulated container, such as a polystyrene cup. The temperature of the solution is measured at the start and end of the experiment to work out the change in temperature. As 1 cm³ of solution is assumed to have a mass of 1 g, a mass of 25 g is used for the mass of the solution.

Energy change = mass × specific heat capacity × change in temperature

$$= 25 \text{ g} \times 4.18 \text{ J/g/°C} \times 18 \text{ °C}$$

$$= 1881 \text{ J or } 1.881 \text{ kJ}$$

Calorimetry can also be used to measure the energy change when solids dissolve in water and in neutralisation reactions.

❓ Test Yourself

1. What happens to bonds in chemical reactions?

2. What is used to measure temperature?

3. Why must the temperature of the water be recorded at the beginning and at the end of the experiment?

4. What sort of energy is stored in food and fuels?

5. What sort of energy is released when food and fuels are burned?

⭐ Stretch Yourself

1. A sample of sodium hydroxide pellets is dissolved in 25.0 g of water. The initial temperature of the water was 18 °C and the final temperature was 35 °C. The specific heat capacity of water = 4.18 J/g/°C.
   a) What sort of material should the container for this reaction be made from?
   b) How much energy is given out in this process? Give your answer in kilojoules to three significant figures.

Practice Questions

Complete these exam-style questions to test your understanding. Check your answers on page 125. You may wish to answer these questions on a separate piece of paper.

1 Copper carbonate has the formula $CuCO_3$.

   a) Complete the equation below to show what happens when copper carbonate is heated.   (1)

      $CuCO_3 \rightarrow CuO +$ ____

   b) What is the percentage of copper in copper carbonate? Show your working.   (2)

2 Calcium carbonate reacts with nitric acid to produce a salt, water and carbon dioxide.

   $CaCO_3(s) + 2HNO_3(aq) \rightarrow Ca(NO_3)_2(aq) + CO_2(g) + H_2O(l)$

   a) Explain what the symbol (aq) means.   (1)

   b) A student reacts 2.5 g of calcium carbonate with excess 2.0 mol dm^{-3} nitric acid.

     i) Calculate the number of moles of calcium carbonate that has reacted. Show your working.   (2)

     ii) Calculate the volume of carbon dioxide gas, in dm^3, produced in this reaction.   (1)

     iii) Calculate the volume, in cm^3, of nitric acid required to react with all the calcium carbonate. Show your working.   (2)

3 Hydrated magnesium sulfate, $MgSO_4.7H_2O$ is known as Epsom salts. A sample of Epsom salts was heated until all the water was removed. It was found that 1.05 g of water was removed by heating, leaving 1.00 g of anhydrous magnesium sulfate.

   a) Calculate the number of moles of water removed. Give your answer to three decimal places. Show your working.   (2)

   b) Calculate the number of moles of anhydrous magnesium sulfate left. Give your answer to three decimal places. Show your working.   (2)

4 Chlorine can be made by reacting hydrochloric acid with a solution of bleach.

   $2HCl(aq) + NaClO(aq) \rightarrow Cl_2(aq) + NaCl(aq) + H_2O(l)$.

   a) 0.12 dm^3 of chlorine was produced in this reaction. How many moles of chlorine was produced? Show your working.   (2)

b) How many moles of hydrochloric acid would be required to produce 0.12 dm^3 of chlorine? Show your working. (2)

....

c) 25 cm^3 of acid was used in this reaction. What was the concentration of the hydrochloric acid? Show your working. (2)

....

5 Barium reacts with water to form barium hydroxide and hydrogen.

$Ba(s) + 2H_2O(l) \rightarrow Ba(OH)_2(aq) + H_2(g)$
0.69 g of barium was used in this reaction.

a) How many moles of barium were added to the water? Show your working. (2)

....

b) Calculate the volume, in dm^3, of hydrogen produced in this reaction. (1)

....

c) Calculate the number of moles of water that would be required to react with all the barium metal. (1)

....

6 A sample of an alcohol was analysed and found to contain 60.0 per cent carbon, 26.7 per cent oxygen and 13.3 per cent hydrogen.

a) Calculate the empirical formula of this alcohol. Show your working. (2)

....

....

b) The relative formula mass of this alcohol was found to be 60. What is the molecular formula of the alcohol? (1)

....

c) The alcohol is one of two possible isomers. Draw the structures of these two isomers below. (2)

How well did you do?

| 0–11 | Try again | 12–18 | Getting there | 19–24 | Good work | 25–28 | Excellent! |

Answers

Atoms and Materials

Pages 4–5 Atomic Structure
Test Yourself Answers
1. Protons and neutrons.
2. Electrons.
3. Charge +1, mass 1 amu.
4. Charge −1, mass negligible.

Stretch Yourself Answers
1. As both elements are in group 2 of the periodic table they have a similar electron configuration; they both have two electrons in their outer shell.

Pages 6–7 Atoms and the Periodic Table
Test Yourself Answers
1. To check them against new evidence that is found.
2. They are in order of increasing atomic number.
3. Periods and groups.

Stretch Yourself Answers
1. Dalton predicted that atoms cannot be divided into simpler substances. It is now known that atoms are made of protons, neutrons and electrons. All atoms of the same element are the same. Scientists now know about isotopes. These are different forms of the same element that have the same number of protons and a different number of neutrons.

Pages 8–9 The Periodic Table
Test Yourself Answers
1. The periodic table.
2. It had not been discovered.

Stretch Yourself Answers
1. 2+.

Pages 10–11 Chemical Reactions and Atoms
Test Yourself Answers
1. Atoms can be joined by sharing electrons or by giving and taking electrons.
2. Sodium and chromium.
3. It consists of two hydrogen atoms and one oxygen atom.
4. It consists of sodium atoms, nitrogen atoms and oxygen atoms in the ratio 1 : 1 : 3.

Stretch Yourself Answers
1. a) KCl.
   b) NaBr.

Pages 12–13 Balancing Equations
Test Yourself Answers
1. Atoms cannot be created or destroyed during chemical reactions.
2. $2Na + Cl_2 \rightarrow 2NaCl$.
3. $H_2 + Cl_2 \rightarrow 2HCl$.
4. It is a liquid.

Stretch Yourself Answers
1. The reacting ions collide together and react very quickly.

Pages 14–15 Ionic and Covalent Bonding
Test Yourself Answers
1. Ions are atoms or groups of atoms with a charge.
2. Electrons are transferred.
3. Shared pairs of electrons.
4. 1+.

Stretch Yourself Answers
1. a) Covalent bonding, with a double covalent bond between the two oxygen atoms.
   b) Ionic bonding between the positively charged sodium ions and the negatively charged chloride ions.

Pages 16–17 Ionic and Covalent Structures
Test Yourself Answers
1. a) Ionic compound.
   b) Giant covalent structure.
   c) Giant covalent structure.

Stretch Yourself Answers
1. Sodium chloride contains sodium ions and chloride ions. The ions cannot move when it is solid, so solid sodium chloride does not conduct electricity. When the sodium chloride is dissolved in water the ions can move, so aqueous sodium chloride does conduct electricity.

Pages 18–19 Group 7
Test Yourself Answers
1. Halogens.
2. Halide ions are formed when a halogen atom gains an electron.
3. Ionic compound.

Stretch Yourself Answers
1. Chlorine + Potassium Iodide → Iodine + Potassium Chloride
   Cl_2 + 2KI → I_2 + 2KCl

Pages 20–21 New Materials
Test Yourself Answers
1. Smart materials have one or more property that responds to changes in the environment.
2. 1 nm–100 nm.
3. C_{60}.
4. In sea spray.

Stretch Yourself Answers
1. Photochromic materials change colour when exposed to bright light.

Pages 22–23 Synthesis
Test Yourself Answers
1. Little automation is possible, at least initially.
2. To let the medical profession know the benefits of the new medicine and why they should consider giving it to their patients.
3. Chromatography.

Stretch Yourself Answers
1. The chemicals are made all the time. Raw materials are continuously added and new products are removed.

Answers to Practice Questions
1. a) Two crosses drawn in the outer shell.
   b) It has two electrons in its outer shell.
   c) Magnesium reacts with a bromine molecule to form magnesium bromide. The ratio of magnesium ions to bromide ions in magnesium bromide is 1 : 2.
   d) Magnesium bromide is an ionic compound. The ions cannot move when it is solid. The ions can move when it is molten.
2. a) The nitrogen atom is bonded to three hydrogen atoms by single covalent bonds.
   b) There are only weak forces of attraction between ammonia molecules at room temperature, which are easily overcome.
3. a) Electrons.
   b) i) The nucleus.
      ii) Protons and neutrons.

Number of Protons	Number of Neutrons	Electron Structure
8	8	2, 6
8	10	2, 6

4. a) Covalent bonding.
   b) There are only weak forces of attraction between methane molecules. These weak forces of attraction are easily overcome.

5. 1 or 2 marks awarded when spelling, punctuation and grammar are weak and the answer is poorly organised.

   3 or 4 marks awarded when spelling, punctuation and grammar has some errors, there is some structure and organisation to the answer but confusion over technical terms outlined below.

   5 or 6 marks awarded when spelling, punctuation and grammar is very good, and there is an organised structure to the answer including the technical terms outlined below:
   - Nanoparticles have unique properties because of the very precise way in which the atoms are arranged.
   - Scientists have found that many materials behave differently on such a small scale.
   - Nanoparticles have a very high surface area to volume ratio, are lightweight, hard and strong.
   - Nanoparticles could be used in new computers/sunscreens/deodorants/drug delivery systems/as better catalysts.
   - They could be dangerous to people.
   - They are an exceptionally small size.
   - Nanoparticles may pass into the body in previously unimagined ways and go on to cause health problems.

Earth and Pollution

Pages 26–27 Evolution of the Atmosphere
Test Yourself Answers
1. 21 per cent.
2. Carbon dioxide.
3. It removed carbon dioxide and produced oxygen.

Stretch Yourself Answers
1. Water, methane, ammonia and hydrogen were placed into sterile flasks and exposed to electrical sparks. These reactions produced amino acids. This showed that chemical reactions could produce the building blocks of life.

Pages 28–29 Noble Gases and the Fractional Distillation of Air
Test Yourself Answers
1. CO_2.
2. The different components have different boiling points.
3. Two electrons.

Stretch Yourself Answers
1. It is filtered and cooled so the gases condense. The mixture is then warmed and oxygen and nitrogen are collected.
2. Helium is not flammable.

Pages 30–31 Pollution of the Atmosphere
Test Yourself Answers
1. Sulfur dioxide.
2. It will have a yellow colour.

Stretch Yourself Answers
1. $2CO + 2NO \rightarrow N_2 + 2CO_2$.

Pages 32–33 The Greenhouse Effect and Ozone Depletion
Test Yourself Answers
1. It encourages algae to photosynthesise.
2. Chlorofluorocarbons.
3. CFCs were used as aerosol propellants, coolants and as solvents.

Stretch Yourself Answers
1. $CFCl_3 \rightarrow Cl\cdot + \cdot CFCl_2$
   The chlorine free radical then reacts with ozone, O_3, molecules.
   $Cl\cdot + O_3 \rightarrow ClO\cdot + O_2$
   $ClO\cdot + O \rightarrow Cl\cdot + O_2$

Pages 34–35 Pollution of the Environment
Test Yourself Answers
1. Bauxite.
2. It is transported using oil tankers.
3. The quarries can scar the landscape.

Stretch Yourself Answers
1. It can be recycled into fleece material to make new clothes.

Pages 36–37 Evidence for Plate Tectonics
Test Yourself Answers
1. The crust.
2. Solid.
3. Silicon, oxygen and aluminium.
4. Iron and nickel.

Stretch Yourself Answers
1. The crust and upper mantle.
2. a) The Earth's lithosphere (crust and upper mantle) is split up into about a dozen large plates. Each of these plates moves slowly over the Earth's surface. The movement of the plates is caused by convection currents in the mantle. These currents are caused by the natural radioactive decay of elements deep inside the Earth, which release heat energy.
   b) It comes from the study of seismic waves.

Pages 38–39 Consequences of Plate Tectonics
Test Yourself Answers
1. Earthquakes are caused when plates suddenly move past or over each other, having been restricted and causing stress to build up. The sudden release of this stress results in an earthquake.
2. There are too many factors involved.
3. Earthquakes occur near volcanoes because plates are moving past or over each other, having been restricted and causing stress to build up. The sudden release of this stress results in an earthquake.
4. The oceanic plates are denser.

Stretch Yourself Answers
1. When the Earth's magnetic field reverses, the iron minerals in the solidifying lava line up with the Earth's magnetic field in the new, opposite direction, forming a symmetrical pattern about the ridge.

Page 40 Everyday Chemistry
Test Yourself Answers
1. New substances are made during a chemical reaction, and there is an energy change.
2. To make it more appealing to users.
3. Thermochromic pigments change colour as the temperature changes.

Stretch Yourself Answers
1. Colloids do not separate because the particles are so small that they are fully dispersed throughout the mixture and do not completely settle to the bottom.

Page 41 The Carbon Cycle
Test Yourself Answers
1. Carbon is found in the atmosphere, oceans and in rocks.
2. More fossil fuels are being burned.
3. Coal, oil and gas.
4. By photosynthesis.
5. More carbon dioxide is being released into the atmosphere.

Stretch Yourself Answers
1. There is a balance between plants taking in carbon from the atmosphere during photosynthesis, plants and animals returning carbon to the atmosphere during respiration and fossil fuels releasing carbon dioxide into the atmosphere during combustion (burning).

Answers to Practice Questions
1. a) Plants.
   b) Carbon dioxide.
   c) Ozone.
   d) Harmful.
2. a) CFCs.
   b) Venus or Mars.
   c) (Early) oceans.
   d) Plants evolved.
   e) Nitrogen was also produced by living organisms such as denitrifying bacteria.
3. a) O_2.
   b) i) It is increasing.
      ii) More fossil fuels are being burned.
   c) SO_2.
   d) Damage to statues; Damage to trees.
4. a) i) Mantle.
      ii) Outer core.
      iii) Inner core.
      iv) Crust.

b) Metamorphic rock.
c) It is denser.
d) West coast of South America.
e) One of: earthquake; volcano; tsunami.
f) There are too many factors involved.
5. 1 or 2 marks awarded when spelling, punctuation and grammar are weak and the answer is poorly organised.

3 or 4 marks awarded when spelling, punctuation and grammar has some errors, there is some structure and organisation to the answer but confusion over technical terms outlined below.

5 or 6 marks awarded when spelling, punctuation and grammar is very good, and there is an organised structure to the answer including the technical terms outlined below:
• Recycling saves raw materials.
• Rainforests, where bauxite is mined, are preserved.
• Less energy is required to recycle aluminium.
• Less carbon dioxide is released into the atmosphere with recycling.
• Landfill sites are not filled up as quickly.

Organic Chemistry and Analysis

Pages 44–45 Organic Chemistry 1
Test Yourself Answers
1. Four.
2. 'Saturated' means they contain no double bonds; hydrocarbons contain hydrogen and carbon atoms only.
3. Methane.
4. C_4H_{10}.

Stretch Yourself Answers
1. a) $C_{56}H_{112}$.
   b) $C_{18}H_{38}$.

Pages 46–47 Fuels
Test Yourself Answers
1. It takes millions of years.
2. Hydrogen and carbon.
3. Three from: runny; easy to ignite; have low boiling points; are valuable fuels.

Stretch Yourself Answers
1. The larger the hydrocarbon molecule is, the stronger the forces of attraction between the molecules are, so more energy is required to overcome the force of attraction.

Pages 48–49 Vegetable Oils
Test Yourself Answers
1. Oils are obtained from fruits, seeds and nuts.
2. Biofuels are fuels made from plant materials.
3. Vitamins A and D.
4. Antioxidants stop foods reacting with oxygen.

Stretch Yourself Answers
1. a) Unsaturated.
   b) Plant.
   c) Add bromine water – if the fat decolourises the bromine water it is unsaturated.

Pages 50–51 Plastics
Test Yourself Answers
1. Addition polymerisation.
2. Monomers.
3. Plastic bags and bottles.

Stretch Yourself Answers
1. The diagram has two carbon atoms joined by a double bond. Each carbon atom is also joined by single bonds to two fluorine atoms.
2. Unsaturated – there is a double bond between the two carbon atoms.
3.

Pages 52–53 Ethanol
Test Yourself Answers
1. C_2H_5OH.
2. Ethanol contains oxygen as well as hydrogen and carbon.
3. Sugar beet and sugar cane.
4. Yeast.

Stretch Yourself Answers
1. If the temperature falls too low, the yeast becomes less active and the rate of the reaction slows down. If the temperature rises too high, the yeast is denatured and stops working altogether.

Pages 54–55 Organic Chemistry 2
Test Yourself Answers
1. Hydroxyl.
2. C_3H_8O.
3. Add universal indicator, which will turn orange–red.
4. $C_nH_{2n+1}OH$.

Stretch Yourself Answers
1. Carbon dioxide and water vapour.

Pages 56–57 Analysis
Test Yourself Answers
1. Modern methods are faster, more sensitive, more accurate, and smaller samples are needed.
2. Chromatography is used to separate out the components of a mixture.
3. Infra-red spectroscopy identifies the type of bonds present in organic compounds.
4. It identifies bonds but all the members of a homologous series have the same functional group.

Stretch Yourself Answers
1. Green $= \dfrac{6.0 \text{ cm}}{8.0 \text{ cm}} = 0.75$

   Blue $= \dfrac{5.0 \text{ cm}}{8.0 \text{ cm}} = 0.625$

Pages 58–59 Cosmetics
Test Yourself Answers
1. Solvent.
2. Solute.
3. Solution.
4. They are cheaper than natural ingredients.

Stretch Yourself Answers
1. The aftershave particles evaporate because, although there are strong forces of attraction within aftershave molecules, there are weaker forces of attraction between the aftershave molecules. When the aftershave is put on the skin, some of the molecules gain enough energy to evaporate. The aftershave molecules can then travel by diffusion through the air and be smelt, even from the other side of the room.
2. Perfumes must evaporate easily to they can travel through the air and be smelt.

Answers to Practice Questions
1 a) Emulsions.
   b) Emulsifiers. One end of the molecule is attracted to oils/fats; the other end is attracted to water.
2. a) C_3H_8.
   b) Yes. It contains carbon and hydrogen only.
   c) Heat is being used to break down large molecules into simpler substances.
   d) i) C_8H_{18}.
      ii) Ethene.
      iii) It is used to make polythene, plastic, polymers and ethanol.
3. a) No. It contains hydrogen, carbon and oxygen.
   b) Yes. It has no double bonds.
   c) Propane.
   d) Alkene.
   e) Propene.
   f) Alkene.
   g) Butene.

4. 1 or 2 marks awarded when spelling, punctuation and grammar are weak and the answer is poorly organised.
3 or 4 marks awarded when spelling, punctuation and grammar has some errors, there is some structure and organisation to the answer but confusion over technical terms outlined below.
5 or 6 marks awarded when spelling, punctuation and grammar is very good, and there is an organised structure to the answer including the technical terms outlined below:

For
- It is the best way to ensure that products are safe for people to use.
- It allows us to develop new medicines that can save people's lives.
- There are strict controls on how animals are treated and when animals can be used.

Against
- It causes avoidable suffering to animals.
- Animals have not given their consent.
- People may respond very differently to the new materials.

Metals and Tests

Pages 62–63 Metals
Test Yourself Answers
1. Metallic bonding is the attraction between the positive metal ions and the sea of negative, delocalised electrons.
2. They have delocalised electrons that can move.
3. Nickel and titanium.
4. A lot of energy is required to overcome the metallic bonds.

Stretch Yourself Answers
1. Nitinol is used in some dental braces.
2. Superconductors only work below their critical temperatures; at present these temperatures are too low to be readily attainable.

Pages 64–65 Group 1
Test Yourself Answers
1. Lithium, sodium and potassium.
2. One.
3. They have the same outer electron structure.
4. Ionic compounds.

Stretch Yourself Answers
1. a) Potassium + Chlorine → Potassium Chloride
   2K + Cl_2 → 2KCl
   b) $K \rightarrow K^+ + e^-$. Potassium has lost an electron so it is oxidised in this reaction.

Pages 66–67 Extraction of Iron
Test Yourself Answers
1. A more reactive metal takes the place of a less reactive metal.
2. Blue.
3. Iron + Copper (II) Sulfate → Iron (II) Sulfate + Copper

Stretch Yourself Answers
1. Zinc + Copper Sulfate → Zinc Sulfate + Copper
   Zn + $CuSO_4$ → $ZnSO_4$ + Cu
   The zinc is oxidised and the copper is reduced.

Pages 68–69 Iron and Steel
Test Yourself Answers
1. Water and oxygen.
2. Hydrated iron (III) oxide.
3. Alloys are mixtures containing one or more metal.

Stretch Yourself Answers
1. The silver should be alloyed with another metal to make it harder. Pure metals have layers because all the atoms are the same size and these layers can pass over each other. In alloys the atoms are different sizes; this causes disruption of the layers so they cannot pass over each other.

Pages 70–71 Aluminium
Test Yourself Answers
1. An alloy.
2. Al_2O_3.
3. Electrolysis.

Stretch Yourself Answers
1. Bauxite has a very high melting point and the addition of cryolite reduces this temperature and, therefore, the energy needed.

Pages 72–73 – Cars
Test Yourself Answers
1. Acid rain, salt water and moist air all increase the rate of corrosion.
2. Aluminium does not corrode.
3. Water.
4. When it burns, hydrogen only produces water, which is non-polluting.

Stretch Yourself Answers
1. Hydrogen + Oxygen → Water
   $2H_2$ + O_2 → $2H_2O$

Pages 74–75 Transition Metals
Test Yourself Answers
1. The middle section, between Groups 2 and 3.
2. Copper is a good electrical conductor. It does not corrode, is easy to shape and does not react with water.
3. The Haber process.
4. Use the carat scale or the fineness scale.

Stretch Yourself Answers
1. Many transition metals can form ions with different charges.
2. Titanium Chloride + Magnesium → Titanium + Magnesium Chloride
3. Titanium extraction involves many steps and requires a lot of energy.

Pages 76–77 Copper
Test Yourself Answers
1. Electrolysis.
2. The negative electrode.
3. They are stronger.

Stretch Yourself Answers
1. a) $Cu - 2e^- \rightarrow Cu^{2+}$
   b) $Cu^{2+} + 2e^- \rightarrow Cu$

Pages 78–79 Chemical Tests 1
Test Yourself Answers
1. A lighted splint is placed nearby. Hydrogen burns with a squeaky pop.
2. Filtration.

Stretch Yourself Answers
1. First, clean a flame test wire by placing it into the hottest part of a Bunsen flame. Next, dip the end of the wire into water and then into the salt sample. Finally, hold the salt in the hottest part of the flame and observe the colour seen.

Pages 80–81 Chemical Tests 2
Test Yourself Answers
1. Carbon dioxide turns limewater cloudy.
2. Blue.
3. Green.

Stretch Yourself Answers
1. $CuCO_3(s) + 2H^+(aq) \rightarrow Cu^{2+}(aq) + H_2O(l) + CO_2(g)$

Answers to Practice Questions
1. a) Limestone and coke.
   b) Haematite.
   c) Hot air/oxygen.
   d) CO_2.
   e) Reduction.
   f) It reacts with silica impurities to form slag.
2. a) Water and oxygen.
   b) The coating stops water and oxygen reaching the iron.
3. a) The atoms are the same size and have a regular arrangement. (1 mark)
      They should be labelled as iron atoms. (1 mark)
   b) Cast iron. In cast iron the atoms are different sizes so they do not have a regular arrangement and the layers cannot pass easily over each other.
   c) Strong.
   d) The more carbon the harder it is to shape.
4. a) Electrolysis.
   b) Bauxite.
   c) Cryolite.

d) It has a lower melting point and bauxite dissolves in molten cryolite.
e) The negative electrode.
f) Reduction.
g) The positive electrode.
h) Graphite/carbon.

Acids, Bases and Salts

Pages 84–85 Acids and Bases
Test Yourself Answers
1. H^+ ions.
2. OH^- ions.
3. Hydrochloric acid, sulfuric acid and nitric acid are all strong acids. They are completely ionised in water.

Stretch Yourself Answers
1. a) Both reactions would produce carbon dioxide/bubbles would be seen/ the same volume of gas would be made in both experiments.
   b) The reaction involving hydrochloric acid would be faster because it is a strong acid while ethanoic acid is a weak acid.

Page 87 Making Salts
Test Yourself Answers
1. Sodium Sulfate + Water.
2. Neutralisation.
3. Hydrogen, H^+, ions.
4. Nitrates.

Stretch Yourself Answers
1. A salt and water.
2. It is when exactly the right amount of alkali has been added to react with all the acid.

Pages 88–89 Limestone
Test Yourself Answers
1. Calcium carbonate.
2. Sedimentary rock.
3. Liquid rock below the Earth's surface.
4. Salt, limestone and coal.

Stretch Yourself Answers
1. a) $CaCO_3(s) \rightarrow CaO(s) + CO_2(g)$
   b) The gas carbon dioxide is given off, which has mass.

Pages 90–91 Metal Carbonate Reactions
Test Yourself Answers
1. Carbon dioxide.
2. Chlorides.
3. Sulfates.
4. Nitrates.

Stretch Yourself Answers
1. $CuCO_3 + 2HCl \rightarrow CuCl_2 + H_2O + CO_2$

Pages 92–93 The Electrolysis of Sodium Chloride Solution
Test Yourself Answers
1. Groups 1 and 7.
2. It is found in seawater and in underground deposits.
3. It is used in bleach, to sterilise water, in the production of hydrochloric acid and PVC.
4. The ions cannot move.
5. Brine.

Stretch Yourself Answers
1. $Na^+ + e^- \rightarrow Na$
   $2Cl^- - 2e^- \rightarrow Cl_2$
2. One of: high blood pressure; heart disease; stroke.

Pages 94–95 Titrations
Test Yourself Answers
1. Alkalis.
2. To ensure the solutions are fully mixed.
3. Colourless.
4. Universal indicator would not be suitable because it is a mixture of several different indicators and it changes colour continuously.

Stretch Yourself Answers
1. Moles of sodium hydroxide = 0.002
   Moles of hydrochloric acid = 0.002
   Concentration of acid = $0.002 \times \frac{1000}{28.5}$
   0.07 mol dm^{-3}.

Pages 96–97 Water and Solubility
Test Yourself Answers
1. Aquifers are rock formations that contain water.
2. Water softeners remove hardness from hard water so the detergent can work properly.
3. Enzymes are added to help remove stains at low temperatures.

Stretch Yourself Answers
1. Some fabrics can be damaged by washing them in water. Other fabrics can be damaged by stains that do not dissolve in water but will dissolve in the solvents used in dry-cleaning.

Pages 98–99 Hard and Soft Water
Test Yourself Answers
1. About 70 per cent.
2. Magnesium and calcium salts.
3. Sodium carbonate.
4. Calcium or magnesium ions are exchanged for sodium or hydrogen ions.

Stretch Yourself Answers
1. The soft water would form a lather quickly with just a little soap solution. If the temporary hard water sample was boiled and then allowed to cool, the hydrogen carbonate ions in the sample would decompose to form carbonate ions. These ions would react with the calcium or magnesium ions present to form precipitates, leaving the water softer. This sample would now react readily to form lather.
   Boiling would not affect the permanently hard water, which would react with the soap solution to form scum.

Answers to Practice Questions
1. a) 0.05 moles.
   b) Moles of acid = 0.10 moles
      Volume of acid = 100 cm^3
2. a) One of: it looks unattractive; makes appliances less efficient; appliances stop working.
   b) Weak = partially dissociated acid = proton donor.
   c) Vinegar.
   d) Strong acids could damage the kettle.
3. 1 or 2 marks awarded when spelling, punctuation and grammar are weak and the answer is poorly organised.
   3 or 4 marks awarded when spelling, punctuation and grammar has some errors, there is some structure and organisation to the answer but confusion over technical terms outlined below:
   5 or 6 marks awarded when spelling, punctuation and grammar is very good, and there is an organised structure to the answer including the technical terms outlined below:
   Soft water does not contain dissolved calcium or magnesium salts.
   As rainwater falls, carbon dioxide in the atmosphere dissolves in the water and then this solution reacts with calcium/magnesium salts in rocks/soils.
   Advantages of hard water
   • It helps to develop and maintain strong bones and teeth.
   • It helps to protect against heart disease.
   Advantages of soft water
   • Using soft water for cleaning reduces costs because less soap is needed.
   • Soft water does not produce limescale.
4. a) Zinc chloride.
   b) Bubbling/fizzing.
   c) It stops bubbling.
   d) Filtering.
   e) Evaporate the water to crystallise the salt.
5. a) Chlorine.
   b) One of: bleach; to sterilise water; PVC; hydrochloric acid.
   c) Hydrogen.
   d) It is used in the manufacture of margarine.
   e) Sodium hydroxide.
   f) One of: soaps; detergents; rayon; acetate.
6. a) $2NaHCO_3 \rightarrow Na_2CO_3 + CO_2 + H_2O$
   b) Baking powder.

Answers

Calculation and Physical Chemistry

Pages 102–103 Relative Formula Mass and Percentage Composition
Test Yourself Answers
1. It is used to compare the masses of different compounds.
2. By adding together the relative atomic masses of all the atoms in the ratio indicated by the formula.
3. 28.
4. 44.

Stretch Yourself Answers
1. 1 mole.
2. 18 g.

Pages 104–105 Calculating Masses
Test Yourself Answers
1. 36 g.
2. 144 g.

Stretch Yourself Answers
1. 2.8 g.

Pages 106–107 Calculations
Test Yourself Answers
1. 0.5 moles.
2. 1.0 moles.
3. 6×10^{22} atoms.
4. 3×10^{23} atoms.

Stretch Yourself Answers
1. 288 dm^3.
2. 4 g/dm^3.

Pages 108–109 Rates of Reaction
Test Yourself Answers
1. It increases.
2. It increases.
3. It increases.

Stretch Yourself Answers
1. If the temperature is increased the particles move more quickly. This means the particles collide more often and, when they do collide, the collisions have more energy. As more collisions have a level of energy greater than the activation energy, the particles react more quickly.

Pages 110–111 Reversible Reactions
Test Yourself Answers
1. It can proceed in either direction.
2. It is where nothing can enter or leave.
3. It is where the rate of forwards and backwards reactions are the same, so there is no change in the overall concentrations of reactants or products.

Stretch Yourself Answers
1. a) Vanadium (V) oxide.
   b) A higher temperature would give a faster rate of reaction but a lower yield of sulfur trioxide. A lower temperature would give a higher yield of sulfur trioxide but a lower rate of reaction. A moderate temperature gives both a reasonable rate of reaction and a reasonable yield of the product.

Pages 112–113 The Haber Process
Test Yourself Answers
1. Ammonia.
2. It is obtained from natural gas or the cracking of oil fractions.
3. The reaction can go forwards or backwards.

Stretch Yourself Answers
1. React the ammonia with oxygen in the air over a platinum catalyst.
   $4NH_3 + 5O_2 \rightarrow 4NO + 6H_2O$
   Allow the nitrogen oxide to cool, and then react it with water and more oxygen to form nitric acid.
   $4NO + 3O_2 + 2H_2O \rightarrow 4HNO_3$
   Then react the nitric acid with more ammonia to form ammonium nitrate.
   $HNO_3 + NH_3 \rightarrow NH_4NO_3$

Pages 114–115 Exothermic and Endothermic Reactions
Test Yourself Answers
1. Exothermic.
2. It releases lots of energy.
3. Exothermic.

Stretch Yourself Answers
1. The vertical axis should be labelled 'energy' and the horizontal axis should be labelled 'time' or 'reaction pathway'. The reactants should be labelled as methane and oxygen and should be higher than the products, which should be labelled as carbon dioxide and water. The energy change and the activation energy should also be marked.

Pages 116–117 Explaining Energy Changes
Test Yourself Answers
1. Bonds are broken and new bonds are formed.
2. A thermometer.
3. To work out the temperature change.
4. Chemical energy.
5. Thermal energy.

Stretch Yourself Answers
1. a) It should be made from an insulator such as polystyrene.
   b) Energy change = mass × specific heat capacity × change in temperature
   = 25 g × 4.18 J/g/°C × 17 °C
   = 1776.5 J
   = 1.7765 kJ
   = 1.78 kJ

Answers to Practice Questions
1. a) CO_2.
   b) $\dfrac{63.5}{123.5} \times 100 = 51.4\%$
2. a) Aqueous/dissolved in water.
   b) i) $\dfrac{2.5}{100} = 0.025$ moles
      ii) 0.6 dm^3.
      iii) Moles of nitric acid = 0.05 = 25 cm^3
3. a) $\dfrac{1.05}{18} = 0.058$ moles
   b) $\dfrac{1.00}{120} = 0.008$ moles
4. a) $\dfrac{0.12}{24} = 0.005$ moles
   b) $0.005 \times 2 = 0.01$ moles
   c) $0.01 \times \dfrac{1000}{25} = 0.4$ mol dm^{-3}
5. a) $\dfrac{0.69}{137} = 0.005$ moles.
   b) 0.12 dm^3.
   c) 0.01 moles.
6. a) $\dfrac{60}{12}, \dfrac{26.7}{16}, \dfrac{13.3}{1}$ C_3H_8O.
   b) C_3H_8O.
   c)
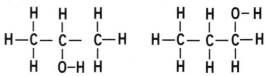

Useful Chemicals

Name	Formula
Ammonia	NH_3
Ammonium sulfate	$(NH_4)_2SO_4$
Barium chloride	$BaCl_2$
Barium sulfate	$BaSO_4$
Bromine	Br_2
Calcium carbonate	$CaCO_3$
Calcium chloride	$CaCl_2$
Calcium hydrogen carbonate	$Ca(HCO_3)_2$
Calcium oxide	CaO
Calcium sulfate	$CaSO_4$
Carbon dioxide	CO_2
Chlorine	Cl_2
Copper (II) carbonate	$CuCO_3$
Copper (II) hydroxide	$Cu(OH)_2$
Copper (II) sulfate	$CuSO_4$
Copper (II) oxide	CuO
Ethanoic acid	CH_3COOH
Ethanol	C_2H_5OH
Glucose	$C_6H_{12}O_6$
Hydrochloric acid	HCl
Hydrogen	H_2
Iodine	I_2
Iron (II) carbonate	$FeCO_3$
Iron (II) chloride	$FeCl_2$
Iron (II) hydroxide	$Fe(OH)_2$
Iron (II) oxide	FeO
Iron (II) sulfate	$FeSO_4$
Iron (III) hydroxide	$Fe(OH)_3$
Lead (II) nitrate	$Pb(NO_3)_2$
Lead (II) sulfate	$PbSO_4$

Name	Formula
Lead (II) iodide	PbI_2
Lithium hydroxide	$LiOH$
Magnesium carbonate	$MgCO_3$
Magnesium chloride	$MgCl_2$
Magnesium oxide	MgO
Magnesium sulfate	$MgSO_4$
Manganese (II) carbonate	$MnCO_3$
Manganese (II) oxide	MnO_2
Nitric acid	HNO_3
Nitrogen	N_2
Oxygen	O_2
Potassium chloride	KCl
Potassium hydroxide	KOH
Potassium iodide	KI
Potassium nitrate	KNO_3
Potassium sulfate	K_2SO_4
Silver nitrate	$AgNO_3$
Sodium carbonate	Na_2CO_3
Sodium chloride	$NaCl$
Sodium hydroxide	$NaOH$
Sodium oxide	Na_2O
Sodium sulfate	Na_2SO_4
Sulfuric acid	H_2SO_4
Tin (II) chloride	$SnCl_2$
Tin (II) sulfate	$SnSO_4$
Water	H_2O
Zinc carbonate	$ZnCO_3$
Zinc oxide	ZnO
Zinc sulfate	$ZnSO_4$

Index

Index

Index